ROUTLEDGE LIBRARY EDITIONS:
URBAN PLANNING

Volume 14

URBAN LAND AND PROPERTY MARKETS IN SWEDEN

URBAN LAND AND PROPERTY MARKETS IN SWEDEN

THOMAS KALBRO AND HANS MATTSSON

Routledge
Taylor & Francis Group
LONDON AND NEW YORK

First published in 1995 by UCL Press

This edition first published in 2018
by Routledge
4 Park Square, Milton Park, Abingdon, Oxon OX14 4RN
605 Third Avenue, New York, NY 10017

Routledge is an imprint of the Taylor & Francis Group, an informa business

British Library Cataloguing in Publication Data
A catalogue record for this book is available from the British Library

ISBN: 978-1-138-49611-8 (Set)
ISBN: 978-1-351-02214-9 (Set) (ebk)
ISBN: 978-1-138-49013-0 (Volume 14) (hbk)
ISBN: 978-1-351-03590-3 (Volume 14) (ebk)

Publisher's Note
The publisher has gone to great lengths to ensure the quality of this reprint but points out that some imperfections in the original copies may be apparent.

Disclaimer
The publisher has made every effort to trace copyright holders and would welcome correspondence from those they have been unable to trace.

Urban land and property markets in Sweden

Thomas Kalbro
Hans Mattsson
Royal Institute of Technology, Stockholm

UCL
PRESS

First published in 1995 by UCL Press.

UCL Press Limited
University College London
Gower Street
London WC1E 6BT

and

1900 Frost Road, Suite 101
Bristol
Pennsylvania 19007-1598

ISBN: 1-85728-052-0 HB

British Library Cataloguing in Publication Data
A catalogue record for this book
is available from the British Library.

Typeset in Palatino.
Printed and bound by
Biddles Ltd, Guildford and King's Lynn, England.

CONTENTS

FOREWORD

As series editors, we are delighted that Thoams Kalbro and Hans Mattsson accepted our invitation to prepare a book on Sweden as volume 5 of the European Urban Land and Property Markets series. The previous four volumes - on the Netherland, German, France and the UK - were all based on research studies undertaken in 1991 for the German Federal Government's Ministry for Regional Town Planning and Building (Bundesministerium für Raumordnung, Bauwesen und Städtbau). All followed the same chapter structure, having been based on the methodology outlined in the Foreword to the Series in each volume. This structure was of course designed to meed a wide variety of national situations.

This volume, based on specially commissioned research carried out by the authors, follows the same specification in terms of the range of themes and topics discussed, but it has been possible to simplify the chapter structure to accord with specific conditions to be found in Sweden.

Richard Williams
Barry Wood
Newcastle upon Tyne
June 1995

Hartmut Dieterich
Dortmund

PREFACE

Describing to an international audience the complex system represented by a country's real estate market is no easy undertaking. On the one hand, authors immersed in the conditions they are describing is liable to overemphasize factors that seem important in a national perspective but are perhaps of less interest to an international readership. On the other hand, they are liable to disregard information taken for granted in the national perspective but which may be fundamental, vital and not readily apparent to people farther afield. Access to the manuscript versions of other books in this series concerning European urban land and property markets has therefore been immensely beneficial. On the whole we have complied with the structure of those books, although at some points modifications have been called for so that Swedish conditions can be more adequately described and explained.

In other respects, too, we have set about our task with some trepidation. The past decade has been a period of great turbulence for Sweden in general and its property market in particular. As will be seen from this book, great changes are taking place in the economy and in regulatory systems. During the period concerned, for example, the property market experienced an unprecedented upturn, only to be plunged into the depths of crisis. Clearly, when this kind of thing happens, parts of the statistical material soon become irrelevant, but we hope that the basic outlines of our description will hold good for some considerable time yet. Another problem is that authors caught up in the midst of dramatic events tend to concentrate on spectacular details, whereas the main contours of developments may elude them. We have been mindful of this problem, but whether or not we have been able to overcome it is another question.

Be this as it may, we have tried to make this book as full of facts as possible, the aim being for anyone interested in Sweden and its property market to have access to relevant information, and in this way to be able to ask pertinent questions and arrive at certain conclusions. At the same time we have tried, as far as possible, to avoid passing value-judgements of our own on developments.

A book of this kind cannot be produced by two authors without the assistance of many able persons whose contributions we wish to acknowledge with gratitude. In addition to the editors of this series, a special word of thanks is due to Dr Andy Thornley, then at the Department of Land Management and Planning, University of Reading, and now Director of Planning Studies at the London School of Economics, for his critical reading of the draft text, whose knowledge of Sweden has enabled him to point out relevant information worth passing on. Mats Bohman and Göran Råckle, economists at the Royal Institute of Technology in Stockholm, have supplied useful comments on certain sections. The case studies would not have been feasible without the kindness and helpfulness of the respective local government officials and company employees. Roger Tanner provided the translation from Swedish into English. Finally, we wish to thank the Swedish Council for Building Research (BFR) for supporting our work financially.

Thomas Kalbro | Hans Mattsson

Stockholm, June 1995

CHAPTER 1
Basic information

1.1 The constitutional and legal framework

Territorial subdivision

The Kingdom of Sweden is, in area, the fifth largest country in Europe after Russia, the Ukraine, France and Spain, but it has only 8.6 million inhabitants. The southern parts of the country are relatively densely populated, whereas the northern parts include extensive and more or less uninhabited regions of mountain and forest.

Sweden is divided into 24 counties (Fig. 1.1), each with its county administration. These are responsible for a large part of national government administration at regional level. But the State is also represented by many authorities at national, regional and local levels. Parallel to the counties (*län*), and more or less co-terminous with them, are 23 county councils (*landsting*). These are headed by elected assemblies and have medical care as their main responsibility. Some of them are also responsible for public transport and various other public services.

At local level, Sweden is divided into 286 municipalities (Fig. 1.1). These vary a great deal in size, as regards both area and population. Largest in area is Kiruna, in the far north, encompassing 19500km^2 and with 26000 residents. Sundbyberg, in the Stockholm region, is smallest, with 9km^2 and 31000 residents. Stockholm is the most heavily populated municipality, with 680000 residents and an area of 190km^2. The Norrland municipality of Bjurholm has the smallest population, with just under 3000 residents in an area of 1300km^2.

Greater Stockholm, which is administered through 22 municipalities, has 1.5 million residents and is the most populous region. The other two metropolitan regions – Göteborg and Malmö – have 700000 and 500000 residents respectively, and are administered at local level through 11 and 9 municipalities.

Thus, the organically continuous metropolitan regions are divided into several municipalities. Otherwise, local administration is based on the principle of a municipality consisting of a central locality and its surrounding area. The surrounding area can include communities of various sizes. In sparsely populated areas the central locality may be small and the surrounding rural area of considerable extent.

National governance

The principle that "all public power emanates from the people" is laid down in the opening section of the Constitution Act (*Regeringsformen*), which is the most important of Sweden's four constitutional statutes. Government by the people is

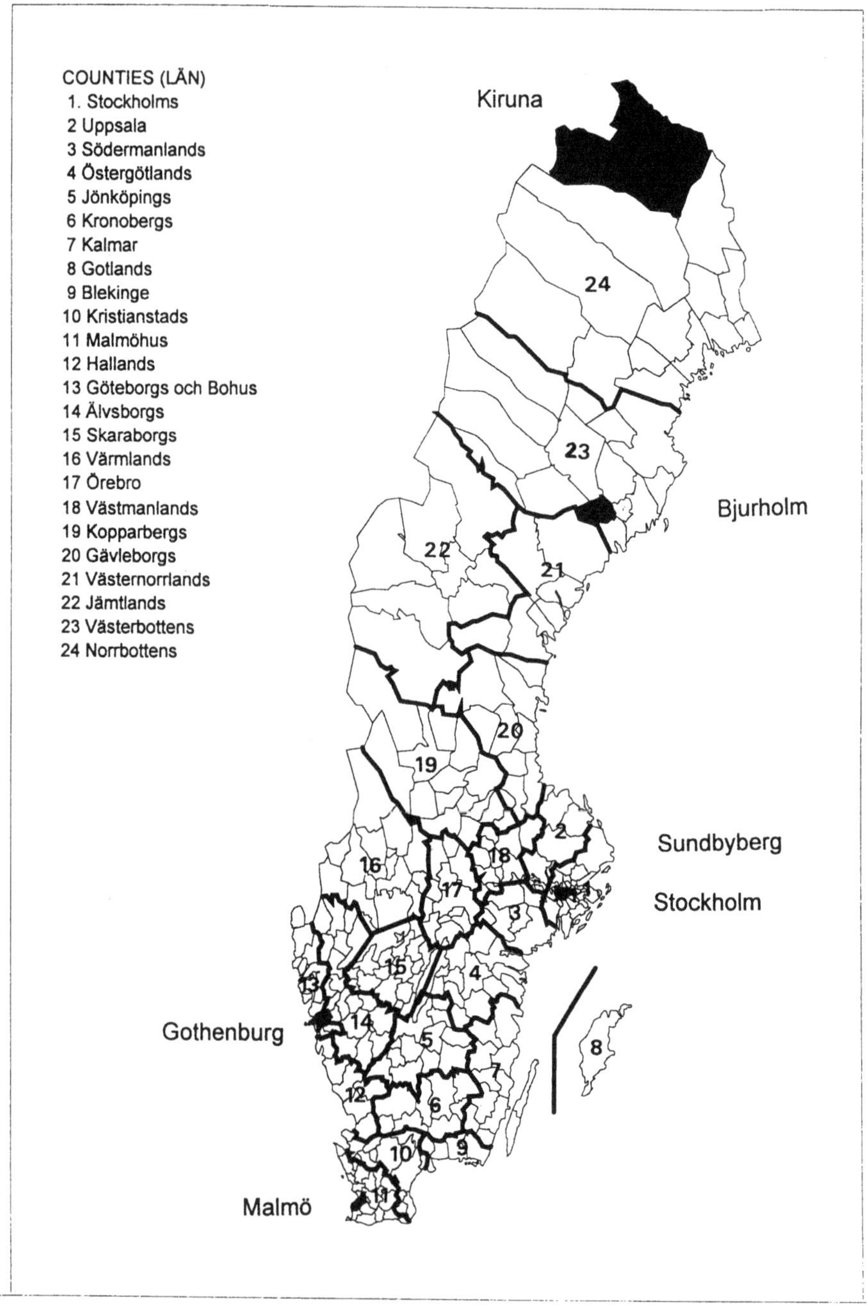

Figure 1.1 County and municipal boundaries in Sweden. Municipalities mentioned in the text are named and shown in black.

indirect and is realized through representative and parliamentary rule and through municipal or local self-government. The principal organs for the public exercise of public power are the *Riksdag* (a single-chamber parliament), the government, administrative authorities, courts of justice and municipalities.

The Riksdag enacts laws, decides national taxation and authorizes the budget. The government (*regeringen*) governs the country within the bounds dictated by its laws and the financial directives of the Riksdag. The government is the planning, initiating and executive organ of State. Its decision-making is collective; that is, no governmental powers are vested in individual ministers. For the drafting of government business, there is a Government Chancery (*regeringskansli*) with ministries (*departement*) for different branches of activity.

Public administration is divided between national, county council and municipal authorities. The national authorities come under the government, except for the Bank of Sweden (*Riksbanken*), which comes under the Riksdag. The government issues more detailed instructions for its subordinate administrative authorities and allocates funding to them, as permitted by the Riksdag's budgeting decisions. However, Sweden's version of "cabinet responsibility" does not allow individual ministers to give orders to the administrative authorities.

Nor can the government dictate decisions by administrative authorities concerning the exercise of authority in relation to a private individual or a municipality, or on the implementation of a particular law. Instead this aspect of government can be influenced through legislation, statutory instruments and general directives. Furthermore, prescriptions by the government may be addressed only to public authorities, not to public servants. The national administrative authorities, then, although under the government, are autonomous in their exercise of power. Finally, neither the Riksdag, nor a municipality nor any other authority can dictate the decision to be made by a particular authority in matters of the kind we have just mentioned.

Justice is administered by courts of law (*domstol*). These are national authorities and are entirely autonomous in their judicial capacity. We may add that it is impossible to draw a hard and fast line between public administration and the administration of justice.

As has already been made clear, administrative duties and the exercise of authority can also be incumbent on a county council or municipality, and may be entrusted, under statutory powers, to private associations (companies, foundations, associations, etc.) and private individuals.

The national authorities can be organized at national, regional and/or local level. The central, nationwide administrative boards (*ämbetsverk*, of which there are about 70 at present) have the task, within their sectors, and under the authority of the government, of preparing and giving effect to policy decisions by the government and Riksdag. The State-owned utilities can have organizations of their own at regional and local levels.

The county administrations (*länsstyrelse*) are the most prominent State authorities at regional level. It is their task to ensure that national objectives in many policy fields have an impact at county level, and also to promote county development. In addition, they are concerned with promoting the co-ordination of national, county council and municipal activities.

The State also owns limited companies, which operate on a commercial basis. Several national authorities and State-owned utilities have recently been turned into limited companies, and some parts of these activities are to be sold off to new interests.

Municipalities and county councils

The municipalities (*kommun*) are responsible for education, except for higher education, which is a State responsibility. They are also responsible for caring services for the elderly, and for physical planning, energy planning, housing supply, fire services, sanitation, and so on. In addition, they are normally responsible for streets, green areas, and water and sewerage in urban communities.

The municipalities are governed by the municipal council (*kommunfullmäktige*), which is politically elected. The principle of municipal self-government means that the municipality is entitled to decide matters of common concern to its residents. The principle also includes the right to levy taxation and charge for municipal services. Thus, an economic foundation exists for autonomous decision-making. The municipalities are at liberty to decide how to organize themselves, except that a municipal executive board (*kommunstyrelse*) is obligatory. In addition the municipality can set up subordinate politically controlled boards (*nämnd*) to which certain decision-making powers are delegated. These boards, including the municipal executive board, are responsible for ensuring that activities are conducted within the boundaries of the goals and guidelines laid down by the municipal council. The boards in turn can delegate matters to individual officials. A board (e.g. a building board – *byggnadsnämnd*), once it has been set up, becomes an administrative authority and as such is autonomous in its particular governmental capacity.

Because the municipality has important and partly autonomous functions in physical and other planning, and is responsible for issuing permits relating to building, an account based on Gustafsson (1992) will now be given of some of the main principles on which municipal administration is based. Municipal policy decisions must be geared to the public interest, in so far as questions (building permit procedures, for example) are not governed by special legislation. However, decisions of public interest need not concern all municipal residents. Furthermore, the decisions must relate to the territory or members of the municipality. Then again, all members of the municipality must be treated equally (the principle of equal status). The municipality may not make retroactive decisions if these are detrimental to its members. The municipality may conduct public activities on a non-profit basis, in order to provide services for its residents. On the other hand, it may not, in principle, engage in speculation, and the charges made for a particular activity must not be aimed at returning a profit (the true cost principle). The municipality may promote enterprise in the locality through general measures such as infrastructure investments, from which all undertakings can benefit on equal terms.

The county councils (*landsting*) are above all responsible for medical care. In some parts of Sweden they are also responsible, in partnership with the municipalities, for regional and local transport undertakings. In terms of administrative law, the county councils are a form of municipality, and they are governed on similar lines. The county councils have independent powers of taxation.

State, county council and municipality are expected to co-operate in the pursuit of common societal objectives, because the constitution lays down that the aim of public activity is to promote the economic and social wellbeing of the individual and, in particular, to safeguard the right to employment, housing and education.

Civic control and the right of appeal

Civic control of the political assemblies - Riksdag, county council and municipality - is ultimately exercised through general elections. Ordinary elections are held every four years, with the possibility of extra ones in between.

Another important element of control over public activity is the constitutional right of citizens (under the Freedom of the Press Ordinance - *tryckfrihetsförordningen*) to inspect public domain documents, i.e. documents kept by national and local authorities. This publicity principle is also one of the preconditions for media surveillance of government, including the government itself, county councils and municipalities. But it can be restricted for certain purposes, such as out of consideration for national defence, general economic interest and personal privacy. To safeguard personal privacy, for example, medical records may not be divulged to unauthorized persons.

Official decisions, including court judgments, can normally be contested by appeal to a higher legal authority. Public administration and the courts usually have a three-tiered decision-making structure, but the actual number of levels of legal authority can be either more or less. The government is the highest authority, or court of final appeal, for several kinds of administrative and political issue. The supreme tribunals are the Supreme Administrative Court (*Regeringsrätten*) for administrative cases and the Supreme Court (*Högsta domstolen*) for all other types of cases. In principle, these tribunals can take cognizance only of cases involving important issues of precedent, and a review dispensation procedure exists to this end. Some specialized tribunals, such as the Labour Court, are the supreme instance or highest authority for certain special types of case.

Legislation and its enforcement

Public power is exercised under the law, a principle laid down in the opening section of the Constitution Act. In principle, laws are passed in general terms and not with reference to individual cases. A law may be amended or repealed only through another law.

Constitutional (*grundlag*) amendments require two identical Riksdag policy decisions with a parliamentary election in between. Other laws (*lag*) are passed by the Riksdag with a simple majority. The government can issue binding ordinances (*förordning*) and also instructions (prescriptions) for administrative authorities. The latter, like municipalities, can in their turn issue binding prescriptions. Ordinances and prescriptions must be kept within the bounds dictated by overriding legislation and, ultimately, the constitution.

The technique for drafting new legislation is worth describing, because the background material to current legislation is often consulted by courts and administrative authorities as an indication of the legislature's intentions. This description is based on Löfmarck (1991).

Legislative activity is normally led by the government. The initiative leading to a new law can take the form of a "*motion*" introduced by one or more MPs. If the initiative is endorsed by the Riksdag, notice to this effect is given to the government, which then has to start an investigation for the drafting of legislative proposals. The ministries too can take the initiative in drafting legislation, sometimes at the instance of a national authority, an organization or some other body. The investigation can be headed by a specially appointed committee or a special investigator. It results in a committee report (*kommittebetänkande*) or a ministerial memorandum (*departementspromemoria*) containing and justifying draft legislation.

The use of committees is a characteristic element of political decision-making in Sweden, and behind almost every government Bill of major importance one finds a report by a committee. The committees are often recruited on a parliamentary basis and can include representatives of interest organizations. This makes them a forum of political compromise. The committee's report or the ministerial memorandum is studied within the ministry and circulated to the appropriate bodies, since the constitution requires necessary information and pronouncements to be obtained from the authorities concerned. Non-governmental organizations and individuals must also be given a hearing to the extent necessary.

The ministry gives consideration to criticism received from the bodies consulted. Important legislative proposals are sent to the Council on Legislation (*lagrådet*, comprising three high court judges) for examination. Following the consultation procedure, the ministry drafts a Bill, setting out statutory proposals and giving reasons for them, which the government then introduces in the Riksdag. The Riksdag then sends the Bill through committee, before deciding whether or not to pass it, possibly with modifications. The Act is then promulgated and published by the government.

New Acts and ordinances from the government are published in the Swedish Statute Book (*Svensk författningssamling*, abbreviated to *SFS*). In addition, an annual digest of the main enactments appears in *Sveriges Rikes Lag*, which is a private publication. Official reports and legislative proposals are published in *Statens offentliga utredningar* (*SOU*) or in ministerial memoranda (*Ds*). Government Bills are published in a special series (*proposition*, abbreviated to *Prop*). Reports by standing committees of the Riksdag, recommending the Riksdag to pass or reject a Bill, are also published, as are motions to the Riksdag.

Needless to say, in their exercise of judicial and public decision-making power, courts and other authorities must comply with current legislation. In cases where the law does not clearly indicate the decision to be made, it is interpreted in the light of the *travaux préparatoires* (preliminary procedures) and of previous judicial decisions, especially by superior courts. However, precedent is not formally binding on inferior tribunals, even though in practice it has considerable impact. The Riksdag frequently employs "paving legislation" (*ramlagstiftning*) in questions of suitability, which leaves scope for interpretation by the authority or political assembly giving effect to the laws. Gradually, however, the scope for interpretation is narrowed down, as new precedents are established by the supreme tribunals. The same goes for administrative decisions by the government.

Rights of ownership over land and personal property

Swedish law distinguishes between real and personal property (*fast* and *lös egendom*), and different rules apply to the two categories in several respects. The following paragraphs describe the main features of rights of landownership; a more detailed account is given in Chapter 3.

Real property is land divided into real-property units or subdivisions (*fastigheter*). These subdivisions include expanses of water, apart from the open sea more than 300m from land and some of the larger lakes. Buildings, conduits and installations (fixtures -*fastighetstillbehör*) are a part of real property. This property can also include easements (*servitut*) and shares in common land (*samfällighet*) and common facilities (*gemensamhetsanläggning*). Conversely, one property may be charged with encumbrances for the benefit of another. It may also be charged with a right of way (*vägrätt*) in favour of the State, with mining rights (*gruvrätt*) and with certain other rights in favour of a particular activity. Water assets (i.e. the actual mass of water) are not included in rights of ownership, even though there is normally a right of extraction from the ground for domestic requirements.

Property units cannot be subdivided horizontally, and so there is no question of different ownership above and below ground or ownership of separate storeys of a building. However, certain forms of "three-dimensional ownership" can be achieved with the aid of easements and uses.

A complete property unit, like shares in a property, can be conveyed by sale, exchange or gift. However, official permission is required in certain situations. Part of the ground surface of a property can also be conveyed. However, the permissibility of the division must be assessed by a special authority (property registration authorities -*fastighetsbildningsmyndighet*), because all changes of property subdivision have to conform with land policy legislation and with any legally binding plans to which the land may be subject.

A property can have more than one owner, in which case the owners have intangible shares in the property. They cannot share the property between them in a tangible sense (e.g. by parcelling) without permissibility assessment. The same applies if heirs to an estate wish to share a property between them in a similar manner.

Theoretically, the owner of a property has the exclusive enjoyment of it within the bounds of current and permissible land use. A change in the manner of use generally requires permission. It is worth mentioning that the State and municipalities own and manage their properties on much the same terms as natural and juristic persons.

Owners of property can restrict their right of use of their own free will by conveying the property or parts of it to another by leasehold (*arrende*), tenancy (*hyra*) or easement. Owners can mortgage (*inteckna*) their property.

The entire property can be conveyed on site leasehold (*tomträtt*) if it is owned by the State, a municipality or some other public body. For the site lessee, site leasehold is equated with ownership of the property in every respect, apart from the payment of an annual ground rent to the landowner. Site leaseholds can be sold and mortgaged. Buildings and other fixtures belong to the site lessee.

Mention must also be made of the tenant-ownership (*bostadsrätt*) concept. Under this arrangement, the property is owned by an economic association. People who participate in this obtain rights of occupancy to a dwelling-unit or to non-housing premises. Ownership of a particular dwelling unit is never possible, since the law does not admit the three-dimensional subdivision of real property.

Individual properties can be expropriated, as can site leaseholds, usufructs and easements. Expropriation (*expropriation*) is unusual, partly because the public sector contains efficient land-negotiating organizations for obtaining the properties or rights that it wants. In addition, certain rights, as well as title to land, can be exacted by means of special legislation; the exaction of a right of way (*vägrätt*) in favour of the State and the acquisition (*inlösen*) of land for a municipal street are two such examples. In some cases the municipality can step into the shoes of a purchaser and pre-empt (*förköp*) a property (see Ch. 3.1).

1.2 The economic framework

Economic developments since 1980

This section is intended to provide a general view of economic developments since 1980, with particular reference to the property market. The dramatic changes occurring in the economy around 1990 should perhaps not be overstated, as this was an outstandingly gloomy phase of perhaps limited long-term significance. On the other hand, one cannot really describe the property market in recent years without reference to fluctuations in the economy generally, and vice versa. We also have to bear in mind that in economic terms Sweden is closely dependent on the world at large, and that economic fluctuations in Sweden accordingly conform quite closely to the international business cycle.

Changes in GNP since 1980 can be seen from Figure 1.2. GNP was SEK530000 million in 1980, rising to SEK1440000 million in 1992 in current prices. Most of the increase is explained by a 130% fall in the value of money. Real GNP growth for the period was 20% or, on average, barely 2% annually. Growth was much the same as for the 1970s, which in turn contrasted starkly with the rapid growth of the 1950s and 1960s, when the annual average was 4%.

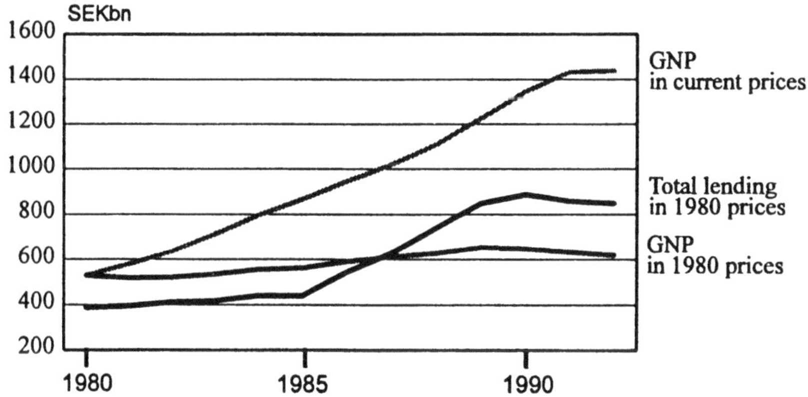

Figure 1.2 GNP development 1980–92 in current and 1980 prices and total lending by credit institutions in 1980 prices (*Sources:* SCB statistical yearbooks, Bankstödsnämnden 1993).

The 1980s opened with a recession. Sweden overcame this problem in the short term by two devaluations in 1981–2, totalling 26% (the same method as was employed during the 1976–7 recession, when there were three devaluations totalling 19%). As a result, the Swedish krona was clearly undervalued for some years. The prolonged international upturn that then followed was beneficial to the Swedish economy. Profits in export industry were also boosted by the devaluation, because there was no reduction of commodity prices. However, the surplus was not ploughed back as investment, and so the outcome was an abundant supply of liquidity in companies and banks. After a couple of years of economic overheating and rapidly rising demand, coupled with a shortage of labour, profits and saving in the business sector were depressed by rapid wage increases.

To understand developments in the 1980s, we also have to consider the deregulation of the credit market. Before 1980, the lending activities of the banks were closely and effectively regulated. A great deal of their lending capacity, as well as the capital that could be attracted through the securities market, was channelled into housing construction and national government borrowing. In this way, industry and the municipalities were to a great extent forced to rely on the international credit market.

In parallel with the regulated credit market, an unregulated finance market emerged within Sweden, funded partly with foreign capital. The advent of this new money market, the inability of the banks to cater for all credit needs within the framework of the regulatory system, and the difficulty of isolating Sweden's capital market from the outside world led to the abolition of credit regulations.

This deregulation was a gradual process. It led to a credit expansion, accelerated by the large increase in the money supply (partly because of the inaction of the Bank of Sweden), inflation and a borrower-friendly taxation system.

Restrictions by the Bank of Sweden on the lending activities of banks, housing societies and finance companies were abolished in 1985. The rule compelling insurance companies to invest in housing bonds and government bonds had been abolished a few years earlier. Property investments abroad became permissible in 1987. The same possibility was opened up for the insurance companies in 1989. Also in 1989, exchange controls were phased out, which meant free flows of capital. Even before that, however, the government's restrictions on international capital flows had to a great extent been rendered inoperative. More than half of the total lending from banks and other credit institutions in this period was to the construction and real estate sector (Bankstödsnämnden 1993).

Institutions that had gained access to new capital as a result of the financial expansion, and which did not have full scope for activity on the international market owing to restrictions on exchange transactions, invested in real estate and securities in Sweden. This helped to raise prices, at the same time as commercial building development grew. After 1987, however, larger foreign investments were feasible and a great deal of money was invested in real estate (and industry) abroad. However, much of the borrowing for these purchases took place internationally.

In addition, there was an interest gap between loans in Swedish and foreign currencies, basically attributable to Sweden's higher rate of inflation, which made it profitable to borrow internationally with a view to investing in the Swedish market. Finance companies old and new, and other financial players, expanded their operations, both borrowing from and competing with the banks. International borrowing grew rapidly. By 1990, Sweden's total net indebtedness abroad exceeded 35% of its total financial assets. In the 1970s, by contrast, international borrowing had been negligible. It is unclear, however, how much of the financial debt is real. Real appreciation of assets abroad is not included in national statistics, and so parts of the debt correspond to real assets abroad.

Household credit grew rapidly after the abolition of the 1985 credit restrictions and, during the ensuing five-year period, households virtually tripled their borrowing. Borrowing was fiscally favoured at the beginning of the period, and inflation made the real rate of interest seem low. The credit bonanza reached its peak in 1988. Subsequently, high rates of interest and limits imposed in 1989 on tax deductibility led to a tailing-off of household credit demand. The net outcome for the period was a large reduction of financial household saving and a growth in private consumption. At the same time, purchasing power was augmented by a high level of public transfers. Thus, a great deal of purchasing power was released, and this had the effect of raising prices. The main emphasis of purchasing was on consumer durables and housing. Housing production grew, and property prices rose. Deregulation and subsidization of housing and building also helped to raise building costs and property prices.

Thus, the fact that the deregulation of the credit market preceded the abolition of exchange controls contributed towards a considerable appreciation of corporate and household assets. Then again, in times of inflation a generous money supply and the ready availability of credit can very easily lead to speculation in real assets. However, the house of cards began to tremble in the autumn of 1990 and it subsequently collapsed. The economy slowed down and went into a downturn that coincided with the international recession. The downturn for Swedish industry, however, set in slightly earlier than in other countries, and appears to be more profound than in most other countries. This was probably attributable to an overcapacity in industry, a cost crisis and a profit collapse resulting from the previous overheating. The structural problems of the public sector, with a rapidly growing State budget deficit and high pressure of taxation, may also have contributed. However, the government was intent on maintaining a fixed exchange rate by unilateral linkage with the ECU. These exertions, in combination with the downturn, led to a drastic fall in the inflation rate between 1991 and 1992. Lower inflation, continuing high rates of interest attributable to external, uncontrollable circumstances, above all in Germany, and the radical tax reform of 1991, which discouraged borrowing and encouraged saving, led to a steep rise in real interest rates after tax (see the expanding gap between the consumer price index and interest on bonds in Fig. 1.3). The high level of interest rates was especially noticeable in the autumn of 1992, when the fixed krona exchange rate was defended by high rates of interest at the Bank of Sweden (500% at one point) and through party political agreements between the governing parties and the opposition on measures for economic recovery. However, in November 1992 the krona was allowed to float, as a result of which it immediately fell by 10–15% against foreign currencies. After four months it had fallen by 20%. This had the effect of resolving a domestic cost crisis attributable to an overvalued currency and boosting export industry. Because of an underlying lack of demand, resulting from a steep growth in household saving, companies catering for the home market were denied similar benefits.

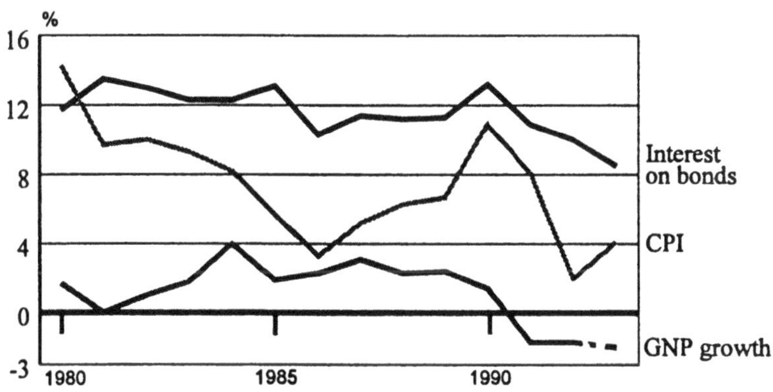

Figure 1.3 GNP rate of growth, average rate of interest on five- and ten-year government bonds and inflation as reflected by changes in the consumer price index (CPI); annual averages, 1980–93 (*Sources:* SCB statistical yearbook, SCB reports serie N, Råckle 1994).

The consequences of economic turbulence

These developments resulted in falling property values, a banking crisis, plummeting share values, growing numbers of bankruptcies and rising unemployment. A government commission (SOU 1993b), appointed to make recommendations on economic policy for the future, describes the property and finance crisis in the following terms. The finance crisis was connected with misplaced expectations on the part of actors in the property and finance markets during the 1980s. Both buyers and lenders apparently counted on a continuing rise in property prices, on the assumption that the boom would continue, resulting in good corporate profits and high demand for office premises. To this were added assumptions of a continuing high rate of inflation and of the taxation system continuing to favour borrowing and investment in real assets, which would mean a low rate of real interest after tax. Furthermore, the deregulation of the credit market made it possible to act on the strength of such expectations, i.e. to borrow money for buying or building property.

The commission observes that, with the shattering of these expectations at the beginning of the 1990s, the property and finance crisis was a fact. Events proved that the State had been in earnest about combating inflation. In addition, the taxation system was recast, and property taxes were raised at a time when the property market was going into decline. To make matters worse, the steep recession led to a drastic decline in demand for offices and other properties. The property and finance crisis, then, seems to have been very much attributable to measures of economic policy to which the actors failed to adjust in time and in some measure could not have foreseen. The crisis was probably also attributable to the inexperience of those involved in the property and finance markets, as regards work in deregulated capital markets, risk assessment and the acquisition of market information. In addition, the sheer mass of commercial building projects started during the boom led to an obvious glut of property once the projects were completed. Vacancy rates increased and rents began to fall.

Despite the slowing down of the economy in 1990, interest rates remained high (see Fig. 1.3). As mentioned earlier, a high real rate of interest coincided with property demand diminishing and the supply increasing. This put a drag on property prices. The ensuing crisis in the construction and property sector affected not only the financial sector itself but also the building supplies industry.

The downturn was reflected by share prices. Post-1980 share quotations for engineering (*verkstäder*), banking (*banker*), and construction and property companies (*fastighets- och byggföretag*) are shown in Figure 1.4. Engineering shares have been included for comparison with the other stocks.

The 1982 devaluation created good prospects for the engineering industry, whose share prices climbing rapidly in 1983. There was a new upsurge in 1986. The international recession of 1987 was also evident on the Stockholm Stock Exchange, as a result of widespread uncertainty about future economic trends. Share prices soon recovered and 1989 turned out to be the peak year. Some way into 1990 share prices fell rapidly before going into a roller coaster routine. In 1993 they had returned to the same level as in previous peak years. This com-

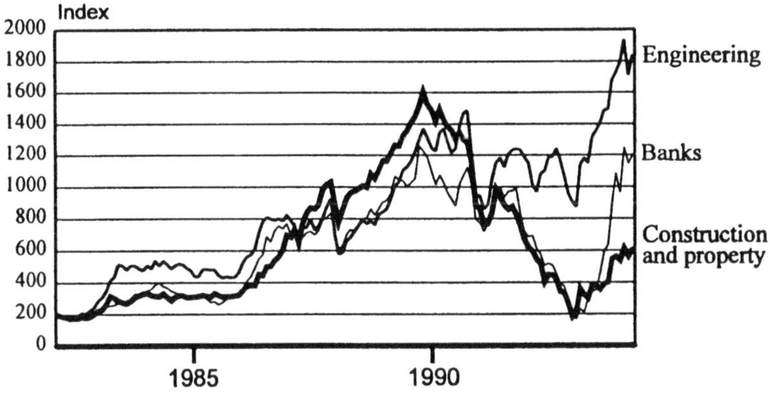

Figure 1.4 Movements of engineering, bank, construction and property company share prices, 1982–93 (1980 = 100; monthly data)(*Sources:* Affärsvärlden 1984–93).

parison makes no allowance for inflation.

Shares in property and construction companies rose in 1983, although not to the same extent as engineering shares. After 1985 they climbed steeply, peaking in 1989. This was followed by a clear decline. The beginning of 1991 brought a temporary resurgence of prices, only to be followed by a heavier drop than ever. By 1992, share values had fallen to a fraction of what they had been three years earlier. Taking a closer look at public property and construction companies between 1989 and 1992, moreover, one finds that some them had disappeared from the stock market, in some cases as a result of bankruptcy.

In the closing years of the 1980s, much of the expansion of bank lending went into property credits. The steep fall in property prices and bankruptcies among property and finance companies had, to say the least, an adverse effect on bank balance sheets, and it is estimated that anything up to three quarters of the heavy credit losses sustained by the banks were directly or indirectly connected with the property crisis. Between 1990 and 1993, credit losses by the banks were around SEK175000 million, and for 1992 alone they were SEK75000 million (Bankstödsnämnden 1993, Ingves 1993). In this way all the leading banks incurred trading losses, although not on quite the same scale as their credit losses. In addition, the State had to underwrite depositors' money in the event of a bank going bust. The crisis also resulted in the banks taking possession of several properties that had been put up as security for credits. The State was also having to take over mortgaged properties. Developments in the property market triggered a banking crisis, as can be quite clearly seen by comparing the movement of property company and bank share prices after 1991 (Fig. 1.4). Bank shares recovered, however, partly perhaps because of the State credit guarantee.

Economic woes were also mirrored by a dramatic rise in the number of bankruptcies, most of them corporate. During the 1980s, bankruptcies were running at between 6000 and 7000 a year. The first signs of an increase came in 1989, and by 1992 the figure exceeded 22000 (only 1000 of which concerned private persons). The greatest increase, in both absolute and percentage terms, came in 1991. Over 13000 of the firms going bankrupt in 1992 had no employees, upwards of 5000 had 1–10 employees, another 1000 or so had up to 50 employees and not quite 200 had more than 50. Thus, it was mainly small businesses that folded (SCB *Reports Serie R*).

One construction company in 20 and also one out of every 20 property management companies went bankrupt during 1992 as well as 1993. In the construction industry, this meant almost 3000 firms each year, with a combined personnel strength of 20000. Such bankruptcies were far less numerous in the 1980s, when the annual rate was about 700 firms with some 2000 employees between them. During 1992 as well as 1993 some 2000 property management companies with a total of 1000 employees went bankrupt (SCB *Reports Serie R*).

Sweden's economic crisis is also reflected in the unemployment figures. All through the 1980s, unemployment fluctuated between 1.5% and 3% of the workforce. In addition, between 3% and 5% did not belong to the regular labour market but were employed through special labour market quality programmes. These programmes include, for example, temporary public employment, special, subsidized job opportunities and employment training. In 1991 there were signs of an impending rise in unemployment as redundancies were announced, jobs were shed, firms went bankrupt and recruitment dwindled. One year later, unemployment skyrocketed, and by mid-1993 overt unemployment was running at 10%. In addition, nearly 5% were employed by means of labour market policy programmes (SCB *Reports Serie AM*).

An image of the future

The aim of economic policy, as expressed in the 1992 Long-Term Survey (SOU 1992), is to secure and enhance the wellbeing of the citizenry. Sub-objectives to this end are a high economic growth rate, full employment, monetary stability, equilibrium in the balance of payments, fair distribution of the standard of living, regional balance and a good environment. As has now been shown, the first three of these sub-objectives came to grief during the latest recession and much work remains to be done in order to rectify the position. In addition, Sweden needs the support of an international upturn. In that case, will there be a repetition of developments following the 1982 recession?

The depreciation of the krona in 1992 seemed one year later to mean a powerful stimulus for the quarter of Swedish enterprise exposed to international competition, whereas sheltered production for the home market, and especially the construction industry, may have some tough years ahead. Can low or non-existent GNP growth be turned into recovery through production increases in industry or through productivity improvements in the service sector?

The shift towards high unemployment is especially disturbing, full employment having long been a top-priority policy objective in Sweden.

The national debt is another worry. For some years it was very low, but apparently it is set to increase drastically over the next ten-year period unless resolute policy measures are taken. The problem is that transfer systems (national health insurance, unemployment insurance, etc.) will then have to be modified, which is likely to meet with strong political objections from various interest groupings. The power of these groups was demonstrated by alternatives suggested by an advisory government commission (SOU 1993b). Another method will be to separate parts of the transfer systems from the national budget.

Given the large volume of housing and non-housing construction in the late 1980s, a decline is expected in building activities. It is partly offset, however, by heavy investments in roads and railways.

Consumption and investments

The following two sections provide further details of the Swedish economy. Some of the tables and figures are merely "snapshots"; the main text has more to say about changes over time.

GNP in 1992 was SEK1440000 million, making SEK165000 per inhabitant. Of this, an average of about SEK90000 went on private consumption. Household assets and disposable income in 1990 are shown in Table 1.1. As can be seen, three quarters of the single-person households with any assets at all owned less than SEK200000 net. Persons in jointly taxed households had more assets, even allowing for the fact that they consist of two adults. Only a small proportion of all households had greater assets.

On the income side, few families had a disposable personal income exceeding SEK300000 (Table 1.1). Most single-person households had less than SEK100000. This group includes many pensioners and young adults. Single parents were better off. Nearly all jointly assessed households had between SEK100000 and 300000. Within this jointly assessed group, families with children had a larger disposable personal income than childless households.

It has been claimed in Sweden that the 1980s were a decade of widening income gaps, but in terms of disposable personal income there is little to support

Table 1.1 (a) Assets (SEK '000)and (b) disposable personal income by household category, 1990 (%). *Source:* SCB, *Statistical yearbook* (1993, own recalculations).

(a) Household category	0.1–200	200–400	400–600	600–1000	1000–	No. of households (000)
Single	75	15	6	3	1	1794
Two-person	40	27	15	12	5	1427
Total						3221

(b)	0–100	100–200	200–300	300–400	400–	No. of households (000)
Single, no children	66	33	1	<1	<1	2674
Single, with children	31	65	4	<1	<1	268
Two-person, no children	8	56	31	4	1	973
Two-person, children	2	26	60	10	2	795
Total						4710

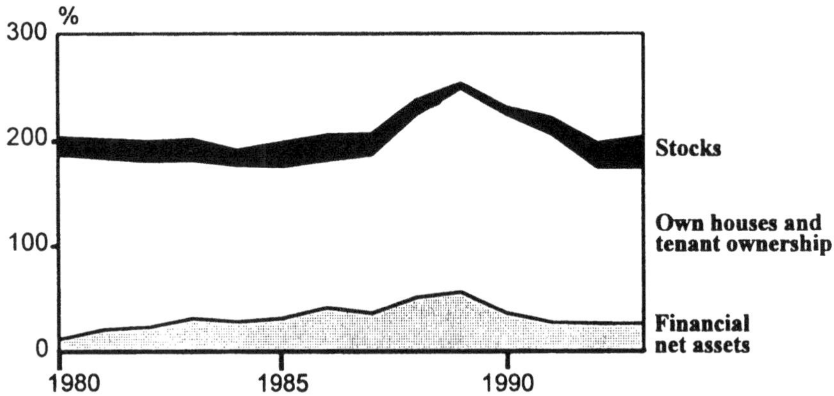

Figure 1.5 Net assets of households in relation to available income, 1980–93 (*Source:* Prop. 1992/3: 150).

this claim. Inequalities in capital assets, on the other hand, were accentuated during the 1980s, and, moreover, capital assets are more unevenly distributed than income. The top 1% of income earners account for 3.5% of the sum total of incomes, whereas the wealthiest 1% own 18% of total net assets (SOU 1992).

Figure 1.5 shows an aggregate estimate of the net wealth of households in relation to disposable personal income and by form of assets. It shows that capital assets are twice the disposable personal income. Mostly they take the form of single-family dwellings and tenant-ownership. The rise in property and share prices during the late 1980s was reflected in a growth of net wealth, notwithstanding a simultaneous rise in debts. The subsequent fall in prices, property prices especially, has not yet been fully offset by a growth in financial saving.

Patterns of household consumption during the 1980s were relatively static in that food, clothing and housing (furniture and equipment included) accounted for almost two thirds of private consumption, and transport (car ownership included) another 15–20%. Foreign travel accounted for about 5%. We may add, while on this subject, that Swedish employees are entitled by law to 25 days' holiday. Much of this time is spent in owner-occupied weekend cottages.

Roughly 20% of consumption during the 1980s was devoted to housing, or 25% if fuel and electricity are included. Housing costs rose in 1991, however, with the result that nearly a quarter of consumption was devoted to housing alone and a further 5% to electricity and fuel (Table 1.2). Increased taxation through an extension of VAT liability was a major cause of this.

The household savings ratio (i.e. savings in relation to disposable personal income) hovered around 2% at the beginning of the 1980s, after which it took a negative turn, bottoming out at minus 5% in 1988/9. The ratio for 1990 was zero, and at the beginning of 1993 it was 8% – a record. These changes would seem by all accounts to be attributable to changes in the taxation system and to households being extra thrifty when times are uncertain. In addition, households presumably need to compensate themselves for the diminishing asset value of homes and securities.

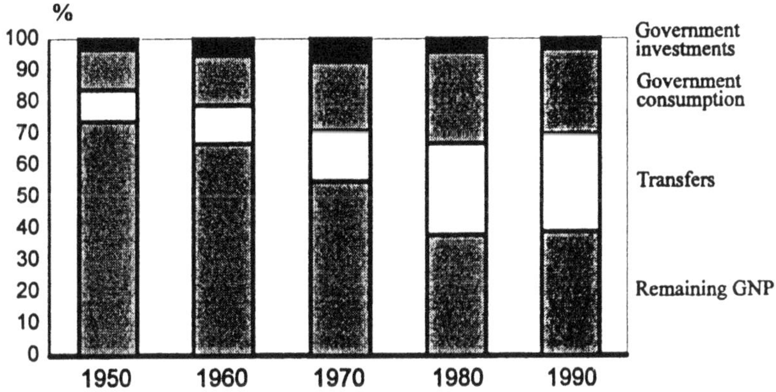

Figure 1.6 Public spending by proportion of GNP, 1950–90 (*Source:* SOU 1993c)

Table 1.2 Breakdown of private consumption, 1991 (%). Source: SCB, *Statistical yearbook* (1992, own recalculations).

Category	%
Food, beverages and tobacco	20
Clothing and footware	6
Gross rent, fuel and power	28
Furniture, furnishing and household equipment and operation	7
Medical care and health expenses	2
Transport and communication	17
Recreation, entertainment, education and cultural services	9
Miscellaneous goods and services	7
Foreign travel	4
Total	100

Public consumption and investments during the 1980s amounted to some 30% of GNP annually, whereas public spending was about 60%. The difference consisted of public transfers. Looking further back in time, one finds that these figures were lower (Fig. 1.6), which shows that transfers in Sweden have increased greatly during the post-war period. Welfare aspirations have made Sweden a high-taxation country. Three quarters of transfers go to households and the remainder to companies, apart from a small proportion allocated for international development assistance. Large transfers are devoted to pensions, health insurance, unemployment benefit, child allowances and housing. The growth of public spending has also been attributable to a growth of public consumption in the form of defence, public administration, the administration of justice and policing, education, medical care and social welfare. The last two items account for nearly half of all public consumption, and also, coupled with education, for the greater part of consumption increases.

Housing transfers in 1992 totalled SEK55000 million, half of which went on direct subsidization of mortgage interest payments and a quarter each on housing allowances and mortgage interest tax reductions (Ministry of Finance 1992).

In addition to the State, municipalities and county councils are also responsible for public activities. The municipalities and county councils have a turnover equalling almost one third of GNP. Of this the municipalities account for two thirds and the county councils for the remainder. Some municipal expenditure takes the form of transfers, but transfer programmes are mainly a national government concern, and so a relatively large proportion of local government turnover is concerned with public consumption.

Investments fluctuated between 18% and 22% of GNP throughout the 1980s, with 1989 and 1990 as peak years. Mainly they comprised private investments, public investments being limited to between 2% and 3% of GNP. Investments and consumption, then, underwent only moderate changes in relation to one another during the 1980s. Small variations between them, however, conceal large percentage changes in the investment item. The economic downturn caused investments to fall by about 1–2% of GNP annually between 1991 and 1993.

Table 1.3 shows investments in proportion to the contributions of different branches of enterprise to GNP production in 1991. Investment-intensive sectors were energy and water supply, transport and communications and real estate, i.e. sectors characterized by a more or less high level of building activity. In fact, more than half of Sweden's total investments went into buildings and facilities, with roughly equal shares for housing and non-housing; 5% referred to streets and roads.

Developments concerning industrial investments merit some attention. After

Table 1.3 Total investments, and investments in buildings and facilities in relation to sectoral contributions to GNP, 1991. Source: SCB, *Statistical yearbook* (1993, own recalculations).

Sector	GNP (SEK bn)	Investments (%)	Of which: buildings and facilties (%)
Agriculture, hunting, forestry and fishing	30700	22	9
Mining and quarrying	4800	23	5
Manufacturing	262400	16	3
Electricity, gas and water	38900	33	22
Construction	98300	6	1
Wholesale and retail trade, restaurants and hotels	126900	14	4
Transport, storage and communication	81800	35	12
Financing, insurance, real estate and business services	295500	38	35
Other private services	50300	11	2
Producers of central and local government services	294700	11	8
Total			

the war and until the beginning of the 1970s, the industrial sector rapidly stepped up its investments in machinery, buildings and facilities. Investments then declined until the end of the 1980s, when they once more expanded in real terms. The post-1970 decline, however, was attributable not to a reduction in plant investments but to a diminishing share of resources being put into buildings and facilities (Industriförbundet 1992).

It is worth mentioning in this connection that, at the beginning of the 1980s, Swedish industry invested heavily in the USA, after which it was Europe's turn. Swedish companies came to invest more abroad than at home, and this was not made up for by foreign investment in Sweden. However, there were large Finnish industrial takeovers in Sweden.

Economic sectors

Commodity production accounted for one third of GNP in 1990. The service sector accounted for two thirds. Public administration alone provided almost a quarter of GNP production. Agriculture and forestry, at 3%, were negligible, although if the food industry is included their contribution rises by a couple of percentage units. Real estate provided one tenth of production (Table 1.4). Business services apart, there were no appreciable changes during the 1980s in the relative contributions of different economic sectors to GNP.

Table 1.4 Sectoral contributions to GNP, 1980, 1985 and 1990 (%). Sources: SCB, *Statistical yearbooks* (various years; own recalculations).

Sector	1980	1985	1990
Agriculture, hunting, forestry and fishing	4	4	3
Mining and quarrying	<1	1	<1
Manufacturing	24	24	22
Electricity, gas and water	3	3	3
Construction	8	6	8
Wholesale and retail trade, restaurants and hotels	12	12	11
Transport, storage and communication	7	6	6
Financing, insurance, and business services	4	8	10
Real estate	9	10	10
Community, social and personal services	4	3	4
Producers government services	25	23	23
Total	100	100	100

Taking industrial output over a longer period of time (the post-war era), however, one finds considerable shifts in manufacturing industry down to 1970, as a result of a change of emphasis from smokestack export industry to engineering. Since then, engineering has provided half the output value of industry. The workforce in manufacturing industry has been roughly 1 million since the 1960s, although in recent years the figure has fallen to 0.8 million. As a comparison, the construction industry employs up to a quarter of a million (SCB Population and Housing Censuses). The number of industrial workplaces has also diminished, especially as regards those with few employees. Developments have favoured the large undertakings. In 1990 the ten largest companies occupied roughly one third of all industrial employees, as against 15% 30 years earlier.

Part of the industrial decline was probably attributable to the devaluations of the 1980s favouring the traditional industries, which have a relatively slow-growing international market. In this way the development of the growth industries was probably impeded. The mainstays of industrial output are motor vehicles, iron and steel, paper and pulp but also high-tech commodities. The pharmaceutical industry has been a dynamic performer.

Because of the small size of Sweden's population, it is greatly dependent on international trade. Exports and imports in 1990 totalled, respectively, SEK340000 million and SEK320000. Engineering provided half the country's commodity exports, forest products one fifth, steel one tenth and chemical products the same amount. The remaining tenth included, for example, energy, food products, textiles and furniture.

The workforce has been declining in industry, agriculture and forestry, while the service sector has expanded. This applies to both public and private services. The public sector had slightly more than 0.5 million employees in 1950, 1 million in 1970 and 1.6 million in 1990. The majority today are employed in social services such as child care, schools, medical care and care of the elderly. As a result of this expansion, one third of the workforce are employed in the public sector. One seventh work in the national government sector and the remainder in municipalities and county councils.

The private services sector has developed on similar lines, with 0.7 million employees in 1950, about 1 million in 1970 and 1.3 million in 1990.

1.3 The spatial, demographic and social framework

The spatial framework

Sweden has a population of 8.6 million and a low average population density: 21 per km^2. There are great regional differences, however, and the southern parts of the country are more than ten times as densely populated as the far north (Fig. 1.7), where there are large uninhabited or sparsely populated mountain and forest regions.

As a result of this low population density, even small communities can have the functions of central localities. Accordingly, in the official statistics, a locality (*tätort*) is defined as a built-up area having at least 200 inhabitants, provided the buildings are not normally more than 200m apart. This definition is exclusively concerned with demographic agglomerations and not with the function of the communities, with the result that suburbs of larger localities are treated as statistical units in their own right, so long as the settlement agglomerations are separated by an undeveloped area of at least 200m. Not all localities, then, have the function of a central locality.

There is no congruency between the statistically defined localities and Sweden's administrative boundaries. The discrepancy can be instanced by the locality and the municipality, respectively, of Stockholm. The locality, which impinges on neighbouring municipalities, has 1040000 inhabitants, whereas the municipal population is 675000. The metropolitan region, with 22 municipali-

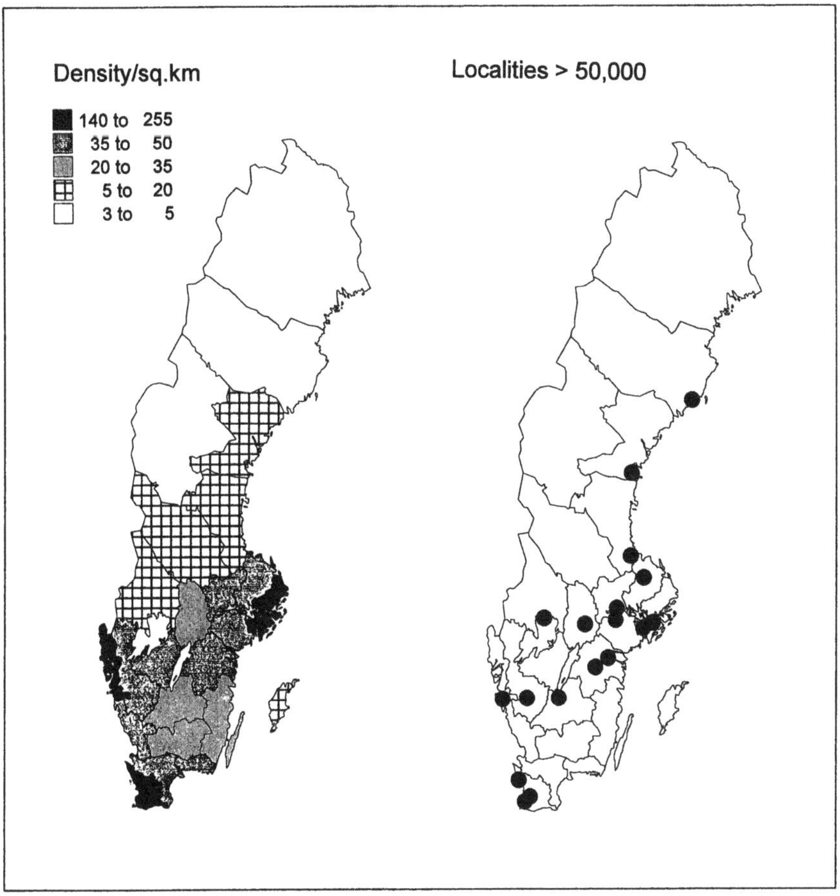

Figure 1.7 County population densities (residents per km^2) and localities with more than 50000 residents (*Sources:* SCB statistical yearbook 1993, SCB 1992j).

ties, has a population of 1.5 million.

Almost 85% of Sweden's inhabitants live in urbanized areas, whereas the remainder live in rural areas (*glesbygd*; Fig. 1.8). About 20% of the national population live in the three largest urban agglomerations: Stockholm, Göteborg and Malmö. Their share of the national population rises to about 30% if one considers their entire metropolitan regions including the suburbs. Each metropolitan region has about 60 or 70 municipalities within it.

Outside the three metropolitan regions there are 15 localities with more than 50000 inhabitants (Fig. 1.7). Rather less than 15% of the national population live in these localities, which are large ones by Swedish standards (Fig. 1.8). If their suburbs are also included, the figure rises to just over 15%. About 90 localities have populations of between 10000 and 50000. At the very bottom of the hierarchy there are more than 1000 localities with fewer than 2000 inhabitants

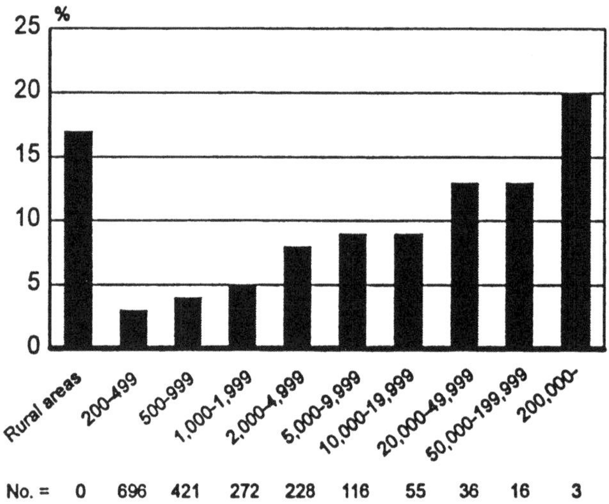

Figure 1.8 Populations of rural areas and localities of various sizes, 1990 (percentage of population and number of localities)(*Source:* SCB 1992j, own recalculations).

each. One tenth of Sweden's population lives in these, the tiniest localities, and about half of them live in localities that, strictly speaking, are suburbs or satellites of larger localities in the vicinity.

One conclusion that can be drawn from the statistical material is that Sweden is a sparsely populated country with mainly small urban concentrations and only a few large agglomerations of population. The "metropolitan cities" are also small by international standards, with regional populations of 1.5 million (Stockholm), 0.7 million (Göteborg) and 0.5 million (Malmö).

There is a great deal of commuter travel between localities. The average Swede travels 40km daily, a quarter of this travel being to and from work. Commuting unites localities, great and small, with each other and with the surrounding countryside to form local and regional job markets. In this way commuting causes networks of localities to form larger employment, service and housing regions than the locality statistics indicate. An approximate picture of urban regions, i.e. of regions with a certain intensity of commuter travel, can be obtained from Figure 1.9, in which circles have been plotted 40km in diameter around the metropolitan cities and 25km in diameter around localities with 15000 or more residents.

Percentage changes in localities between 1980 and 1990 are shown in Table 1.5 (right column). Growth, on average, is highest in the smallest localities, after which it declines with increasing size of locality. However, there is considerable variation within each size category, especially in the smaller localities. Segregating localities respectively within and outside the 40km and 25km circles (as described above) and at the same time grouping them by size, one finds that growth, on average, is fastest in small localities near the metropolitan cities and

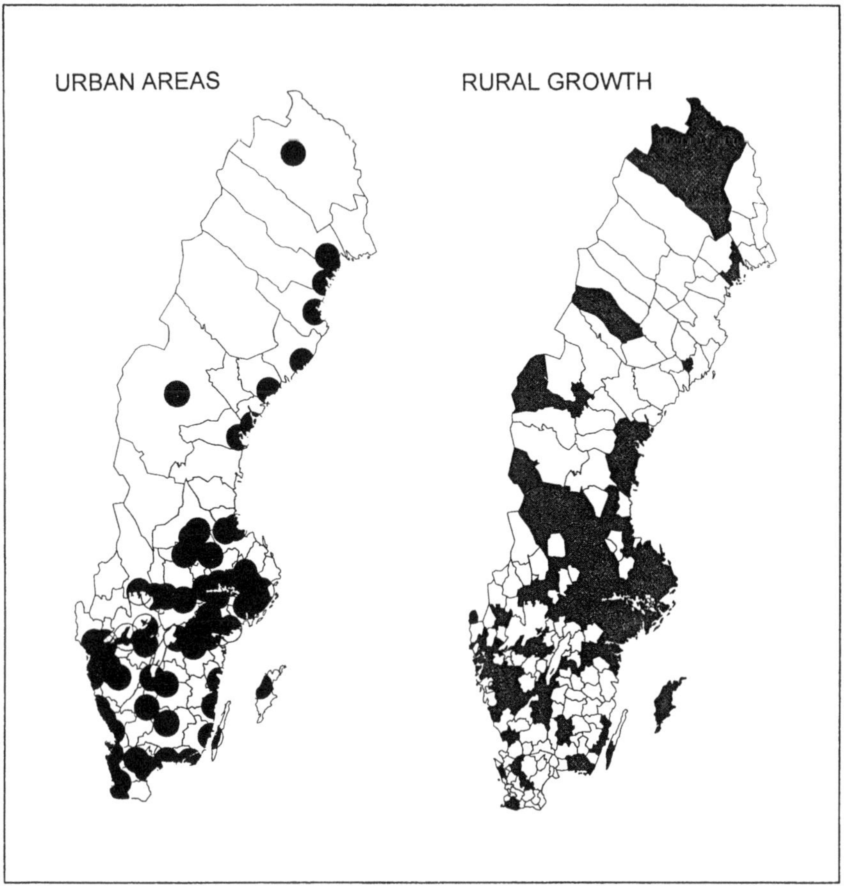

Figure 1.9 Areas within 40km of metropolitan cities and within 25km of other localities with at least 15000 residents in 1990 (black areas on the left-hand map) and population growth 1980–90 in rural areas at municipal level (dark areas on the right-hand map)(*Source:* Nyström 1994).

near other localities with more than 20000 residents; in other words, it is suburban growth, connected with the job market in larger localities. This also implies an outward spread of population towards the suburbs. Another interesting circumstance is that two thirds of all localities come within the circles we are referring to, that is, within what may somewhat loosely be termed urban regions. What is more, roughly 90% of Sweden's urban population comes within these regions.

At the same time as the smallest suburban localities are expanding, rural growth is observable in regions with a reasonably high population base (Fig. 1.9). We may note that a growth area exists in the interior of Norrland (in the middle of the figure), unconnected with the larger urbanized areas. This is a mountain district with highly developed skiing amenities. In the immediate vicin-

Table 1.5 Population changes in localities within and outside "urban regions", 1980–90, and number of localities in 1980. Source: SCB (1992j; own recalculations).

	Size of central locality in 1980							
Size of locality	Stockholm	Göteborg	Malmö	50000–110000	20000–49000	10000–19999	200–9999	Total
(a) Percentage of population change (1980–90)								
200–499	52	31	25	7	16	3	–1	6
500–999	24	50	9	19	9	2	–3	6
1000–1999	48	16	17	12	7	2	–1	5
2000–4999	14	35	5	10	4	1	1	5
5000–9999	5	13	18	3	–2	2	0	3
10000–19999	13	11	6	3	1	1	–	3
20000–49999	12	–	0	–9	0	–	–	2
50000–110000	–1	–	13	2	–	–	–	2
Metropolitan	5	0	–1	–	–	–	–	3
Total	7	5	5	4	2	1	–1	3
(b) Number of localities (1980)								
200–499	10	19	26	81	106	171	265	678
500–999	12	17	11	54	79	88	155	416
1000–1999	5	12	10	38	55	54	96	270
2000–4999	6	8	11	50	35	30	81	221
5000–9999	11	6	10	18	11	20	27	103
10000–19999	4	4	2	6	6	39	0	61
20000–49999	6	0	2	1	22	0	0	31
50000–110000	1	0	1	13	0	0	0	15
Metropolitan	1	1	1	0	0	0	0	3
Total	56	67	74	261	314	402	624	1798

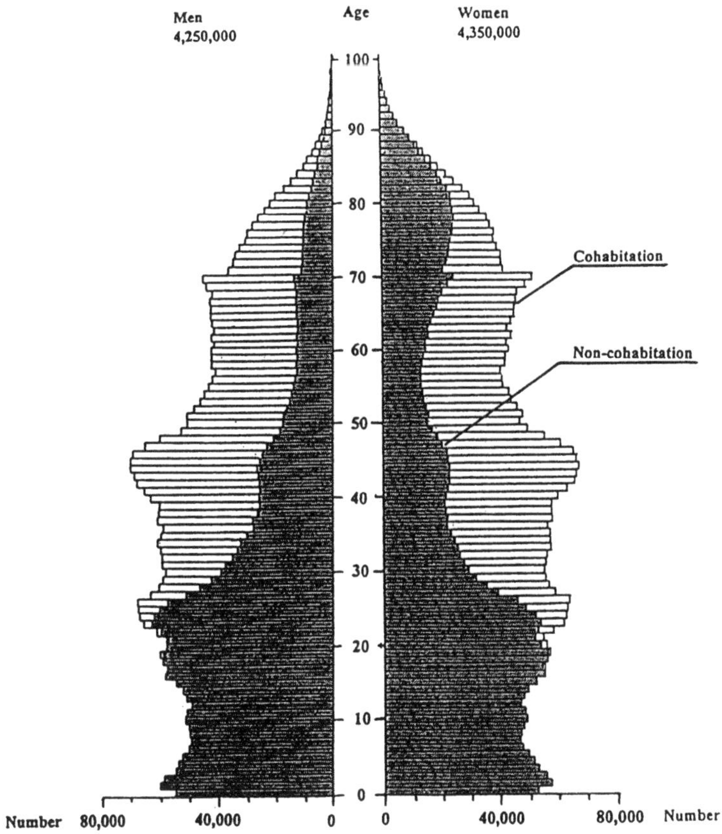

Figure 1.10 Population by sex, age and cohabitation/non-cohabitation, 1990 (*Source:* SCB 1992c).

ity of Malmö, in the extreme southwest, there is no significant growth, despite proximity to a metropolitan city. This is a region of high agricultural productivity, which means firm resistance to the spread of scattered settlement.

Population

The Swedish population pyramid is typical of western European societies in that the balance between different age groups is relatively even up to 70 (Fig. 1.10). The age group born in the 1940s is large, as are the generations of 20–30 year olds and infant children. A large proportion (18%) of the population is above the official retiring age of 65. In fact, Sweden has the oldest population of any industrialized country.

Reproduction rates are low. For 1970, 1980 and 1990, respectively, they were 0.9, 0.8 and 1.0. During the period, in other words, the population did not reproduce itself. In fact no generation of Swedes has reproduced itself at any time in the twentieth century.

Minorities

Population growth, however, is positive, because of an excess of immigration over emigration. This has been the case throughout the post-war era. Economic expansion after the Second World War generated labour demand in both industry and the public sector. Labour immigration continued some way into the 1970s, after which it receded heavily and, at the end of the 1980s, was superseded by asylum-seekers and their next-of-kin (SCB 1991c).

The earlier immigrants were rapidly integrated with the Swedish labour market and, after a time, achieved both a higher employment participation rate and higher average earnings than the native population. The opposite applies to refugees and even when they have been living in Sweden for a long time there are great differences between their employment participation rate and earnings and those of the majority population.

More than 820000 of Sweden's 8.6 million inhabitants were born in some other country (SCB *Statistical Yearbook 1993*). Of these 320000 come from other Nordic countries, 230000 from the rest of Europe and 260000 from other continents. Where individual countries are concerned, immigration from Finland (210000 persons) predominates.

Native linguistic minorities with mother tongues other than Swedish exist in the north of Sweden (40000 Finnish speakers and 9000 Sami/Lapps).

Household structure and employment participation rate

Changes in household structure can make a considerable difference to the property market, especially as regards rental and single-family properties. During the past 20 years the number of households has increased and their size has diminished. In 1970 there were 3.1 million households comprising, on average, 2.6 persons each. The corresponding figures in 1990 were 3.8 million households and 2.1 persons. The proportion of single- and two-person households was upwards of 50% in 1970 and about 70% in 1992. There are few households comprising five or more persons (Table 1.6). The three metropolitan municipalities have most small households, half of them single-person households.

Table 1.6 Households by number of members, 1970, 1980, 1990 (%). Source: SCB (1992e).

	1	2	3	4	5+	Total
1970	25	30	19	16	10	100
1980	33	31	15	15	6	100
1990	40	31	12	12	5	100

The main reason for this growth in the number of small households is the growth of the elderly population. The proportion of households consisting entirely of over 65s increased by about 70% between 1970 and 1990 and now represents a quarter of all households in the country. Another reason for the increase, probably, is that single-person households are now easier to establish than they used to be, thanks to the increased housing stock. Younger persons no longer have to continue living at home as long as they used to. About 40% of men and 20% of women aged between 20 and 24, however, live with their parents, and so the proportion of small households can be expected to go on increasing. On the other hand, these age groups are likely to form new families within the next five or ten years, which reduces the number of households (see Fig. 1.10). Divorces, on the other hand, help to increase the number.

The employment participation rates of men and women are roughly equal. For persons aged between 20 and 65 it is 80–90%. The employment participation rate, in fact, is so high that economically active persons have never before had so few inactive persons to support. At least, this was the case until unemployment began rising.

Mobility

The national relocation rate has been relatively stable for the past 100 years, with about 7–9% of the population relocating annually. The average Swede moves house ten times in his or her life, and eight or nine of these relocations involve distances of less than 50km. Most relocations occur for housing and family reasons - young people move away from home; couples move in together or separate. Relocation, of course, can also be attributable to a change in the size of the family or to new economic circumstances (SNA 1991).

Long-distance relocations, i.e. more than 50km, occur, on average, only once or twice in a lifetime. These relocations are commonest among people who have been studying for a long time. One obstacle to long-distance relocation is that about 95% of married men and some 90% of married women between the ages of 25 and 55 are gainfully employed, and finding employment simultaneously for two adults in a new area can be difficult. It is true that persons over 55 show a declining employment participation rate, but not many people in the higher age groups are prepared to relocate. In fact, residential mobility is already declining after age 35. Migratory flows are for the most part two-directional between different areas, and as a rule the numbers of persons moving, respectively, into and out of a community are roughly equal. Before 1970, however, there was a distinct net migration from the countryside to towns and cities. A trend inflection then occurred, bringing a stabilization of population except for the most sparsely populated areas, where the decline continued. The large-scale tendency prevailing after 1970 was for net migration during the 1970s to go from large communities outwards, whereas the 1980s witnessed a slight in-migration surplus in the metropolitan cities. At the same time, younger persons tended to be somewhat more often attracted to larger than to smaller communities, whereas the opposite applies to families with children (SNA 1991).

Tendencies of change

Economic conditions and values are changing all the time, and this leaves its mark on the pattern of settlement. The following, partly contradictory factors may in future come to play an important part:

- The job market is changing as a result of high unemployment. New workplaces may come to be established in different regions from those where there is unemployment, which will lead to migratory movements. But there is also a risk of high unemployment persisting in certain regions.
- As Sweden is located on the fringe of Europe, companies tend to choose locations in the southern parts of the country in order to be nearer to the European mainland. This can result in southward population movements.
- If national boundaries in relation to Europe are opened up for a free exchange of goods, services and employment, competition for existing home markets will presumably increase. Matters are not likely to become easier for operations already exposed to competition. Education, therefore, will be a vital factor in this competition. For this reason technology, for example, and also other branches of education, will be given greater priority in the higher education system. This might strengthen Swedish industry, but since highly educated persons are more disposed to migrate, it could also result in a stream of migrants to the metropolitan cities and the rest of Europe.
- Greater environmental awareness on the part of the population may steer migration in the direction of attractive areas. This can result in a movement of young persons to localities in connection with education and first-time employment, whereas young families move to smaller localities and rural areas when their children arrive.
- A high female employment participation rate tends to reduce the residential mobility of families.

1.4 Land, property and construction

Land use, property stock and value of real property

Sweden comprises 410000km^2 land and 40000km^2 water, sea areas excluded. The greater part, about two thirds, is devoted to production, mainly forestry. Not quite one tenth, 36000km^2, is farmed. The remainder mostly comprises non-productive land (*impediment*) of different kinds. A little more than 2% (11000km^2) of the total land area is incorporated in urban areas, and of this area localities occupy roughly 1% (5100km^2), whereas buildings and installation outside localities occupy another 1%. The latter group comprises land for agricultural buildings, permanent and second homes, building plots including roads and transport operators land, power lines, industry, mining, and so on (Table 1.7).

Localities are often sparsely populated and nearly always include large green spaces. "Greenness" is clearly apparent from Table 1.8, which shows that only six-tenths of all locality land is built on, half of it with housing. The true percentage of built-up land is smaller, because our figure includes gardens and other

Table 1.7 Land-use categories, 1990. Source: SCB (1993h).

Category	km^2	%
Forest land	243300	54
Agricultural land	35200	8
Built-up land	11100	2
Rock surface and mountains	71200	16
Marsh, bog	49500	11
Water	39000	9
Total	450000	100

Table 1.8 Land use in localities, 1985. Sources: SCB (1992g, 1990c; own recalculations).

Land use	km^2	%
Housing	1740	34
Industries	460	9
Communications	600	12
Offices, public service, recreation areas	410	8
Agriculture	300	6
Forestry	740	15
Non-productive land (*impediment*)	790	15
Water	60	1
Total	5100	100

land for buildings and ancillary uses.

Compared with larger localities, the proportion of built-up land in small localities is small and the incidence of detached houses high. Although nearly all localities make a "green" impression, the proximity of nature is especially noticeable in small localities. Then again, a single-family dwelling, on average, takes up 10 or 15 times as much land as a rental apartment (1800m^2 as against 130m^2).

In principle, all land is divided into real-property units. The property stock will be described in more detail when we come to consider the property market (in Ch. 5), but a brief outline at this point will not be out of place. There are no comprehensive statistics concerning the number of property units, but the stock can be estimated indirectly, thanks to the taxation of real property. A tax assessment unit normally corresponds to a property unit, which can also comprise several property units or a functional part of one unit. In 20 out of Sweden's 24 counties the property stock is known and the discrepancy between the number of assessment units and the number of property units is about 1% (authors' calculations based on material from SCB 1992i, CFD 1993). In some counties and for certain property categories, the discrepancies may be greater, but on the whole the assessment units agree with property subdivisions, and so, by consulting the assessment figures, one can arrive at a rough estimate of the number of property units in the country as a whole, as well as their ownership, use and aggregate value.

The structure of ownership will be dealt with in Chapter 5, with reference to the different property categories. Briefly, though, the greater part of Sweden's real-property capital, and of its total land area, is owned privately or by legal bodies. The State, however, owns large forest areas and especially non-productive land including nearly all mountain regions. Furthermore, the State, municipalities and county councils own many tax-exempt units such as hospitals and schools, and the greater part of all infrastructure such as public highways, municipal streets, railways and water and sewerage networks. Finally, the State and certain municipalities own large parts of the power industry, either directly or through companies.

Altogether there are some three million assessment units (*taxeringsenhet*; Table 1.9). Half of them are single- and two-family dwellings (*småhus*) for permanent habitation and another fifth are second homes (*fritidsbostad*). There are upwards of 100000 multi-dwelling and commercial units (*hyreshus*). This group includes both dwellings and buildings with non-residential premises. The number of agricultural units (*lantbruk*), including forestry units (*skogsbruk*), is just over 300000.

Table 1.9 Assessment units by use. Number of units, total assessed values and estimated current value, 1992. Sources: SCB (1993a,e,i; own recalculations).

Property type	No. of units	Assessed values	Estimated values
Agriculture (incl. forestry) with buildings	229000	190	250
Agriculture (incl. forestry) without buildings	94700	60	100
One- and two-dwelling buildings	1538200	590	1100
Multi-dwelling and commercial buildings	112000	430	1200
Industry, electrical power, pit, etc. units	64700	290	600
Secondary homes	584600	90	200
Sites and land for development	181100	25	50
Units worth less than SEK1000	131300	–	–
Tax-exempt units	95700	–	–
Total	3031300	1675	3500

As can also be seen from Table 1.9, the combined assessed value is SEK1700 billion. Assessed values, however, are set at 75% of estimated market value two years before assessment. If the assessed values are written up by means of purchase price coefficients (P/A = purchase price of properties sold/assessed value of the unit), the total property value comes to about SEK3500 billion in 1992 prices. This, of course, is a tentative assessment, and it does not measure the value of Sweden's total capital of buildings and installations, because infrastructure, such as roads, railways and harbours, is not included in these calculations. Besides, the term "tax-exempt units" (*skattefri enhet*) often conceals large facilities, for the most part publicly owned, such as wastewater purification plants, central heating plants, hospitals, schools and churches (see also Ch. 5.8).

To the above estimates must be added Statistics Sweden's (SCB 1992h) estimate of national wealth. The total value of building and construction was estimated at upwards of SEK2600 billion for 1990, and of this total one fifth referred to infrastructure, mainly facilities for transport and for the distribution of electricity, gas and water. Land and other non-reproducible assets were valued at upwards of SEK1500 billion, of which SEK240 billion referred to agricultural and forestry land and rather less than SEK1300 billion to other land. Thus, the combined value of land, buildings and installations would seem to be SEK4100 billion, as compared with the estimated total national wealth of SEK4700 billion.

The above calculations suggest that the value of real property is roughly three times GNP.

Investments in building and construction

In Chapter 1.2 we saw that Sweden's total investments equal one fifth of GNP output. More than half these investments were in buildings and installations. During the 1991–3 recession, however, investments declined heavily by about 10% in fixed prices per year.

Table 1.10 shows building investments broken down into the categories of enterprise, housing and public authorities, and also as a percentage of GNP output. It is clear from this table that the 1982 devaluation was not followed by a surge of industrial investment in real property, even though the devaluation favoured Swedish industry. Investment did not rise until the end of the decade, and the downturn made itself felt in 1990.

Table 1.10 Building and construction investments as a percentage of GNP, 1980–91. Source: SCB, *Statistical yearbook* (1992) and SCB (1993a).

Year	Enterprise	Housing	Public	Total
1980	5.2	4.6	2.7	12.5
1981	4.7	4.5	2.7	11.9
1982	4.6	4.3	2.5	11.3
1983	4.5	4.1	2.3	10.9
1984	4.6	4.2	2.1	10.9
1985	4.5	4.4	1.9	10.8
1986	4.3	4.1	1.7	10.2
1987	4.4	4.5	1.6	10.5
1988	4.4	5.0	1.7	11.0
1989	4.8	5.5	1.8	12.1
1990	4.6	6.0	1.8	12.4
1991	3.8	6.5	1.6	11.9

Public investment in building and construction declined steadily throughout the 1980s, except for the closing years of the decade. The municipalities, on average, accounted for more than half of these investments, the State for a quarter and the county councils for the remainder, apart from a small percentage for the established Church of Sweden.

For many years now, housing construction has been a major investment item. In relative terms, housing investment declined in the early 1980s, but the end of the decade brought an unmistakable expansion. Investment fell off again after 1992, and especially in 1993. This investment item includes both new production and reconstruction of older permanent homes. At the beginning of the 1980s the decline in the production of new homes was offset by reconstruction. The change of emphasis was encouraged by generous State improvement loans. Refurbishment activities continued, however, when new construction picked up again, with the result that total housing investment expanded during the second half of the 1980s. By 1990 it had reached almost SEK50 billion. More than three quarters of the reconstruction resources were devoted to multi-family dwellings, the remainder to detached or semi-detached houses.

There is a heavy concentration of building activities within the metropolitan regions. These accounted for a total of 46% in 1988, about 25% being in Greater Stockholm, 13% in Göteborg and 8% in Malmö. We may note, by way of comparison, that these regions have 32% of the national population (SOU 1990c).

Production costs and cost movements

Between 1966 and 1975, the political target was the production of 1 million homes. In the course of this "Million Programme", real production costs were reduced by rapid efficiency improvements in a production-orientated building industry. This was made possible partly by large-scale construction projects and the use of new materials and techniques. Productivity declined after 1975 as a result of the diversion of building operations from new development to renewal, and also to the diminishing size of projects, simultaneously with an alarming rise in material prices. The development of costs was unfavourable compared with industrial products, and still more in relation to consumer goods. One resultant problem was that, for a given housing requirement, an increasing proportion of national resources had to be set aside for housing production.

The movement of housing construction costs post-1980 for multi-family dwellings on undeveloped land is shown in Figure 1.11. Prices, in fixed money terms, were reasonably stable for the first five years. They then started to climb, at the same time as several economic indicators pointed upwards and new output of housing and non-housing was on the increase. The recession of the past few years, on the other hand, has brought a reduction of costs in real prices. Normally, detached and semi-detached housing estates are 10% cheaper per square metre to build than flats in multi-dwelling buildings. New flats in reconstruction areas are on the other hand 15% more expensive on average.

The cost of other building development conforms on the whole to housing production costs, although levels depend on the type of construction project involved (detached or semi-detached houses, non-housing premises, standard of equipment, etc.). Certain variations also occur, depending on the location of the project. Building costs in Greater Stockholm are a good deal higher than in other parts of Sweden.

In addition to the building costs referred to above, total production costs also include the cost of land, foundations and connection charges for streets, water and sewerage. These items normally make up 15–20% of total production costs for multi-family dwellings and 20–30% for detached housing estates. The difference is attributable to the multi-family dwellings having more dwelling units per square metre site area. For non-housing the land percentage can be still greater,

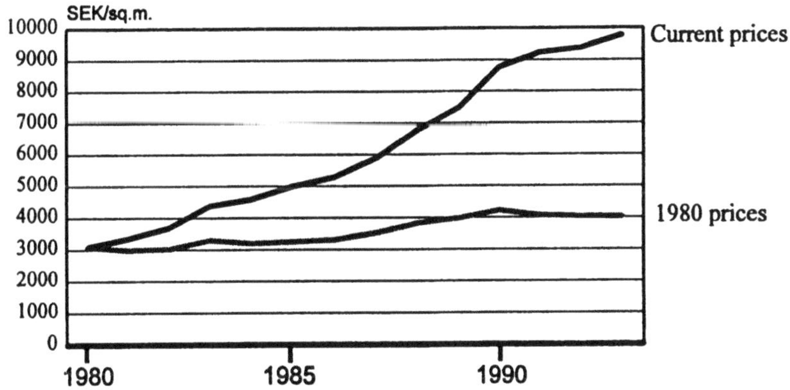

Figure 1.11 Movement of construction costs per m^2 for multi-family dwelling units, 1980–93 (*Sources:* SCB 1993a, 1994b).

in which case it is often attributable to high land prices.

Construction industry

Contractors Building is ordered by developers - that is, the State, county councils, municipalities, business undertakings, private persons and so on, who undertake or commission building, demolition, site works or construction on their own behalf. The building is often done by contractors. Developer and contractor, however, may be identical, or parts of the same enterprise, in which case the enterprise builds under its own auspices.

Just over 230000 people were employed by construction companies and building trade enterprises in 1993. Building is mostly a local activity, and this is reflected in the structure of these enterprises. The market is characterized by the existence of a few large construction enterprises, nearly all of them public companies, and many small concerns. According to Byggentreprenörerna (1994, figures refer to 1991) there are about 7000 contracting firms (for house building, heavy engineering and earth-moving). About 6000 of these firms have fewer than 20 employees. These smaller firms often figure as subcontractors to a "main contractor". For example, contractors' machinery services, using earth-moving machinery, loaders, excavators, and so on, are provided mainly by small, independent businesses competing mostly within small, local markets. Often these undertakings collaborate within local contracting co-operatives, some of which have dominant positions in the local market.

The smaller undertakings account for over 30% of all turnover. The number of building trade enterprises specializing in electrical installation, ventilation, heating and plumbing, painting, and so on can be estimated at more than 10000. Both the construction and building trade sector also include many "one-man firms" with no employees.

There are also about 20 large businesses with over 500 employees, accounting between them for nearly 50% of all turnover in the sector. The three largest contractors, Skanska, NCC and BPA, have 50000–60000 employees. The smaller and larger firms have essentially different markets. The smaller firms derive nearly half their turnover from repairs, alterations and enlargements. The larger ones, by contrast, devote 80–90% of their activities to the construction of new buildings and installations.

As regards corporate development after 1970, it is above all the medium businesses that have been reduced. The largest, on the other hand, have grown, very much by means of take-overs and mergers. Structural change in the contracting sector has resulted in the two largest construction enterprises developing into nationwide organizations for both housing and non-housing production. These companies have also aimed to achieve vertical integration of their operations by dealing with subcontractors, building materials, management of large property stocks of their own and financial administration. Property management has most often materialized through the development and upgrading by companies of building projects of their own that they have subsequently retained. In addition to the two largest companies there are about 30 undertakings that occupy strong positions in regional markets, and about ten of these operate virtually nationwide. Very few foreign contractors are active in Sweden, and their operations are confined to works requiring special technical skills and competence (SOU 1990c).

It is common for building enterprises to collaborate on a consortium basis for major construction projects, to facilitate competitive tendering. Consortia can also be formed for smaller projects.

Building materials Building materials account for roughly one third of construction costs. This market consists of innumerable submarkets, and here again there is heavy concentration. Several subsectors have fewer than four production enterprises. This applies, for example, to the cement, light concrete, mineral wool and building slab markets. Other sectors with more producers are still in many cases dominated by one or a few companies. This applies, for example, to flat glass, doors and flooring. This heavy concentration is probably attributable to the poor profitability of building materials in industry during the late 1970s and early 1980s, which prompted rationalization mergers. Another reason may be efforts on the part of these companies to prepare for a growth of international competition (SOU 1990c).

Property management

Property management employed about 40000 persons in 1988 and is expected to employ 60000 at the turn of the century (SOU 1990a). The main activities from an employment viewpoint are housing management, management agency (*uppdragsförvaltning*) and estate agency services. Altogether property management is roughly estimated by Lundström (1992) to involve 150 million m^2 multi-family housing, 75 million m^2 offices and commercial premises, 75 million m^2 industrial facilities and 90 million m^2 publicly owned premises – the State 15 million, county councils 25 million and municipalities 50 million m^2.

Most properties are managed by their owners, especially where large holdings are involved. The organizational structure of property management, however, appears to be changing, with property owners tending more and more to transfer management duties to specialized companies.

CHAPTER 2
The policy environment

This chapter scans the history of land policy and urban planning over the past 50 years. The reason for taking this period is that the existing legal, economic and organizational order is rooted in the social policy framed after the Second World War and consolidated between then and the mid-1970s.

The underlying ideology continues to dominate present-day Swedish thinking on the subject of planning. But the building activities of the 1980s and the official inquiries of the 1990s suggest that urban development is coming to be seen in a new light.

2.1 General land policy

Swedish regional policy may seem contradictory in certain respects, perhaps because it is partly a result of political compromises but also because it takes into account social consideration for disadvantaged groups. The discrepancies are apparent, for example, from the fact that cities and central localities are viewed in a positive sense as growth centres, at the same time as public supportive measures are deployed to sustain rural communities, not least in the interior of Norrland.

At the beginning of the 1970s, for example, the Riksdag adopted a plan for building up a system of localities, based on strong, differentiated central localities with robust local job markets. The intention was for these localities to counteract the drift to the really large communities, while supporting in various ways the drift from the land. One means of building up the system of localities was for State investments to be concentrated on the designated central localities. Side by side with this central locality approach, supportive measures were developed for the economic life of rural communities, the aim being for them also to be kept alive. At present the locality priority principle seems to carry less weight than rural policy.

Regional policy has a bearing on land policy, but national land use is more profoundly affected by agricultural and forestry policies. Extensive support has been given to farmers so that they can achieve reasonable incomes, at the same time as the State has committed heavy resources to rationalizing agriculture and reducing the number of persons actively employed in it. Clearly, the importance of agriculture in the national economy has been declining. Forestry has meant more, and great pains have been taken with an active forestry policy aimed at keeping forest-based industry supplied with raw materials. In recent years, though, the pursuit of rational production on forest land has had to be balanced against demands from conservancy and environmental interests. Then again, agricultural and forestry policies encroach on one another, because of farmers often being forest owners as well, to a greater or lesser extent.

Another, economically marginal, but area-demanding activity that at times comes into conflict with conservancy and forestry interests is reindeer herding. The reindeer pasturage rights of the Sami (Lapps) take legal precedence over roughly one third of Sweden's land area, within both State-owned and other properties. These grazing lands are located in the far-flung forest and mountain regions of the north.

The State also controls and supervises other land uses besides agriculture and forestry. To a great extent this is done through sectoral policy and with the assistance of central and local authorities. In addition, the State is directly responsible for the construction and management of public highways other than municipal streets in urban areas, railways, shipping lanes, large airports and military training grounds. The State has extensive interests in power production facilities, power lines and telecommunications and is actively involved in areas concerned with social and scientific nature conservancy, heritage conservation and mineral extraction. Special State organizations supervise the environment, and court orders are required for certain types of environmental interference. Furthermore, the State follows up and issues both mandatory provisions and recommendations on urban development. A large part of all property subdivision is effected by the State.

The municipalities also play a central part in the achievement of State targets. One way or another, they participate in nearly all changes of land use, even those for which the State is responsible. State and municipality are expected to co-operate. In addition, the municipalities have been made directly responsible for several land-related questions, such as housing supply, physical planning, refuse collection and disposal, water and sewerage in localities and street maintenance in localities. The municipalities work independently within the framework of their financial resources. When State and municipality come into conflict with each other and are unable to arrive at a consensus solution to their problems, the outcome will depend on which of them, by law, has the last word in the matter concerned.

The small municipalities existing previously were successively amalgamated into larger units to make them economically viable. Urban and rural municipalities were amalgamated as well. The main purpose of these boundary changes was to achieve rational units for the conduct of social welfare services and, later, for the running of schools. Until 1952 there were 2500 municipalities, but by 1974 the number had been reduced to 278. Since then a few municipalities have been re-partitioned. Although these amalgamations were not primarily intended to create strong municipalities for the conduct of land and housing policy or physical planning, opportunities in this respect were radically improved by amalgamation. As time went on, the municipalities were made increasingly responsible for the planning of housing production and for physical planning. After 1974, for example, the State required them to draw up comprehensive, although not legally binding, physical plans for each municipality in its entirety.

In addition to the State having influenced land use for a long time through sectoral policy and through the municipalities, during the mid-1960s the government embarked on a more comprehensive form of "national physical planning" (*fysisk riksplanering*), aimed at more closely controlling the localization of heavy, environmentally disruptive industry and urban development. Among other things, the national politicians were dissatisfied because, in reality, they did not have full powers of assessment regarding large-scale industrial projects; instead those projects came up for assessment only at a late stage when, after municipal assessment, most things were already cut and dried.

The outcome of this State commitment to national physical planning was perhaps rather meagre. Developments had to some extent outdistanced the idea of being able to control physical development through nationwide plans (Forsberg 1992). Start-ups of heavy industry were mostly a thing of the past. The population drift from the countryside to urban communities began to ebb, and the building of second homes declined from its previously high level. Besides, the municipalities became large enough and strong enough to take a more independent line vis-à-vis the State. Not that physical planning was entirely without effect. Sensitive areas were protected from second-home development, and that protection was indeed codified as part of the 1987 reform of planning legislation. In addition, the importance of comprehensive planning was brought home to the newly formed "large municipalities".

Land legislation was revolutionized during a short period around about 1970. The Riksdag replaced old legislation by passing a new Property Code, Property Subdivision Act, Environment Protection Act, Water and Sewerage Act, Tenancy Act and Roads Act. It passed a Pre-Emption Act and a Housing Renewal Act. Important amendments were made to existing laws, such as building legislation, the Expropriation Act and the Nature Conservancy Act. Many of these changes were prompted by a desire for more effective legislation, but also by a desire to reduce the impact of proprietary interests on matters of land use.

Some of these statutory reforms emanated from policy standpoints defined in connection with work on national physical planning. In 1972, for example, the Riksdag abolished the last remaining opportunities for unrestricted building by making new building development in rural areas subject to approval, which it had not always been previously.

New urban development rights were regulated for in 1948. As a result, all new building development today is subject to municipal approval. An exception is made for agricultural buildings and, since the beginning of the 1980s, for certain small outbuildings. In certain more sensitive areas, such as protected shorelines, the location of new building development is also examined by central government, unless these powers have been delegated to the municipality.

There have also been land legislation reforms in more recent years, the most important of them perhaps being the 1987 Planning and Building Act and the

Conservation of Natural Resources Act, adopted at the same time. These enactments followed very prolonged official inquiries and extensively codified parts of the planning practice that had evolved previously. The new Acts were not very radical but could rather be seen as a continuation or perhaps even a rounding-off of the sweeping changes of the 1970s. They included certain novelties, such as the above-mentioned codification of areas for protection, the requirement of environmental impact assessments in certain connections, wider civic participation, stipulated implementation times for detailed development plans and clearer compensatory provisions for encroachment situations.

2.2 Policy for the environment

Views on environmental pollution changed rapidly during the 1950s and 1960s when it became clear that something would have to be done about industrial pollution. Provisions were codified in 1969 in a new Act, the Environment Protection Act, for the protection of the external environment. This Act established safeguards against water pollution, air pollution, and noise and other forms of pollution. A new authority, the National Licensing Board for Environment Protection (*koncessionsnämnden för miljöskydd*), was set up, with the task of carrying out prior assessments of environmentally hazardous activities, especially large-scale industrial and wastewater emissions. The environment protection work that has been done since then has very substantially improved the water of lakes and water-courses and the immediate environments of industrial facilities.

Gradually it became clear that environmental problems were not always attributable to quantitatively large emissions but also involved individual substances with very severe environmental impact. All chemical products in use have therefore been made subject to a system of chemical control regarding their effects on health and the environment.

The emphasis of environmental work has shifted more and more from the rectification of past transgressions to the prevention of new ones. Successive enactments have supported this by stressing the conditions for long-term ecological, social and economic interests in connection with urban expansion. A government committee has proposed gathering all environmental legislation into a separate code, to underline the importance of environmental issues (SOU 1993c, 1994). It is evident, though, that several environmental problems are of a global nature, and must therefore be solved by agreement with other countries, through the adoption of international conventions.

2.3 Housing policy

The evolution of housing policy down to the mid-1970s

Urban land policy, being a matter of prime interest in this book, will be described at greater length. The most central position here is occupied by housing policy, which has more or less permeated all other urban land policy in the past 50 years.

Urban land-use regulation in the modern sense began in 1874, and in the early years of the twentieth century a succession of instruments was used for controlling the expansion of towns and quasi-urban communities, mainly out of consideration for public health.

The embryo of a specific policy on housing had already been created before and during the First World War, but it was not until the 1930s that radical ideas on housing, social participation, the environment, and so on evolved and achieved some impact on housing development. Inputs were selective, concentrating on disadvantaged groups such as large families (SOU 1945).

After the Second World War the Riksdag laid down guidelines for a more comprehensive social housing policy, aimed at achieving good housing conditions for the entire population. The former selective housing policy was now turned into a general policy that would include the population as a whole and not just the underprivileged. Housing came to be looked on as a social right.

The more exact housing policy target defined by the Riksdag was the provision of healthy, spacious, practical and well equipped homes of good quality, at reasonable cost, for the entire population. In addition, encouragement was to be given to non-profit housing management. That is, the structure of ownership was to be changed from private to public or collective. Gradually these aims were operationalized through the definition in quantitative terms of such qualities as healthy, spacious and well equipped.

As a result of the new housing policy, housing production and the housing market came to be almost completely directed and controlled by the public sector. This change, however, did not come as an instant radical reform. Instead it resulted from successive changes to regulatory provisions and housing finance. The aims, however, were clear from the very outset, and for a long time the essential outlines were supported by a political consensus.

In pursuit of these aims the Riksdag modified regulatory provisions, public organizations and conditions of finance. It laid down that both State and municipality were to take part in the implementation process, with the State responsible mainly for regulatory provisions and the efficient supply of resources, whereas the task of the municipalities would be to ensure that building took place. In this way housing supply was made a municipal responsibility. First and foremost that responsibility was incurred by the large municipalities, but gradually it devolved on all municipal authorities.

The government introduced subsidized national housing loans and grants and defined cost ceilings for the award of State support. The structure of loans and grants has changed over the years but the principle of paramount State responsibility has remained. Most of the newly produced dwellings came to be State-financed (by the end of the 1980s nearly 100% of new housing production was financed in this way). In addition, a large share of the assets of banks, insurance companies and, later on, the General Pension Funds was channelled into the housing market through mandatory holdings of housing bonds.

The State and the municipalities also built up a system of individual support for underprivileged groups, so as to bring sufficiently large and well equipped dwellings within their reach.

New building legislation was passed in 1948. Planning was looked on as a municipal concern, and the municipalities were empowered to decide when, where and how urban areas could be built up, in that they were entitled to decide for or against the adoption of legally binding detailed plans. The landowner, however, had the right of production. That is, he decided whether or not he wanted to build. But the production problem was usually solved by the municipality being actively involved in the planning and construction process and, frequently, having land of its own.

The State did not relinquish control of planning and building completely but up until 1987 ratified the municipally adopted detailed plans, that is until the new Planning and Building Act transferred all decision-making powers to the municipalities. The State also had the last word on subsidized State housing loans and issued detailed regulations on the design of housing and other buildings. Therefore, one can say that most things were regulated down to the smallest detail. Regulation was further facilitated by the standardization of building components and fittings by municipal and other developers and by the construction industry.

As we have now seen, the municipalities were required to improve and guarantee housing standards and to distribute State loans and grants for housing construction. They were to provide long-term housing supply programmes and organize housing allocation services. The municipalities were responsible for physical and other planning so as to ensure that housing supply would be sufficient in volume at the right locations and appropriate in terms of types of building and dwelling, forms of tenure and location. In addition, the municipalities were to ensure that the older stock of substandard housing was modernized.

Gradually the municipalities built up large organizations of their own for land procurement, planning, credit supply, development of utilities (mainly water and sewerage), and so on. They set up non-profit housing utilities for the construction and management of rented housing. These utilities are municipal housing companies (*kommunalt allmännyttiga bostadsföretag*) operating without private profit (just under a third of all multi-family dwellings built since 1945 are owned and managed by such utilities). Large municipalities established housing exchanges and, in addition, developed purchasing principles for the acquisition of housing in times of shortage.

Legislation and statutory instruments were gradually amended to divert the profits of building away from property owners and, especially, from those conveying greenfield sites to building interests. There was a danger of State subsidies in the form of loans and grants being fed into land prices. In addition, the large demand for housing also entailed a risk of developers and landlords being able to reap excessive profits on the sale of properties and tenant-owner titles and on the letting of dwellings. This danger also had to be averted.

An active municipal land policy that included land banking was enjoined by the government and Riksdag, and the recommendation was for every municipality to have sufficient reserves of land for ten years' building development. Expropriation facilities were gradually enlarged so that municipalities would be able more easily, well in advance and at low cost, to acquire land for urban development. Pre-emption was introduced in 1968, entitling municipalities in certain situations to take the purchaser's place when land came up for sale.

A Land Ordinance (*markvillkoret*) introduced in 1974 (it was repealed in 1992) stipulated that all land that was to be used for housing development financed by State credits was to be purveyed through the municipality. In this way the municipality would become the sole purchaser of land, renewal properties included, which would have a restraining effect on prices. The Ordinance was seldom observed in practice as exemptions were obtainable (Vedung 1993). The percentage of building on municipal land actually declined after the Ordinance was issued. Its ineffectiveness was mainly attributable to the municipalities being in such a strong position through the planning monopoly that they could get their own way in matters of urban development and so there was little point in land banking. Municipal land purchases, in fixed money terms, had already peaked in 1967, after which they declined steadily (purchases in 1980 being down to one third of the 1967 level; Bladh 1991).

Site leasehold is another land policy instrument originally created so that less affluent persons could build their own homes without having to buy land. Instead the occupant paid an annual ground rent to the municipality (site leasehold conveyance is the prerogative of public landowners). During the 1960s, site leasehold instead had become a means for the municipality to retain the land and, by regularly increasing the ground rents, to benefit from the appreciation in land values.

The use of site leasehold was enjoined by the Riksdag in 1967 and 1974, but the actual decision rested with the municipalities. Quite a few of them made large-scale use of site leaseholds, whereas others were wary of tying up capital in land that would yield a return only in the long term.

In various ways, then, the State dissociated itself from rights of ownership, which was logical considering that the home was looked upon as a social right and not as a commercial commodity to be bought and sold in the market. Thus, the Riksdag indicated preferences for rental tenure and displayed some scepticism about tenant-ownership and even more about freehold home ownership, because these two forms of tenure were saleable. From time to time, influential political voices were also raised in favour of transferring all land to public ownership.

Imbalances in the housing market gave landlords an opportunity to charge high rents. Legislation, however, prevented those rents from rising to market levels, at the same time as steps were taken to counteract a black market for rental homes. Rent controls had already been introduced during the Second World War. Direct regulations were gradually abolished and superseded by the principle of the rents charged by non-profit housing utilities setting the norm for housing rents in general. Freehold houses and, eventually, tenant-owner homes were less strictly regulated and could often be sold at a good profit. Gradually attempts were made to overcome this by fiscal means.

The instruments developed for the implementation of housing policy can be summed up as State credit supports, support for population groups with limited resources, rational housing production, extensive municipal planning, an active municipal land policy and a system of rent restrictions (SOU 1974a).

In spite of all these public efforts, a severe housing shortage prevailed until the beginning of the 1970s, owing to migratory movements, a general rise in living standards and a certain amount of population growth. The situation was especially precarious in the mid-1960s, and so the Riksdag set the target of building 1 million new homes in ten years. This was accomplished by 1975. For both ideological and cost-related reasons, output during the initial phase of this "Million Programme" was dominated by multi-family dwellings.

The outcome of housing policy

The market for flats was already showing signs of saturation in 1972. Output declined and increasing numbers of flats remained vacant. Production of detached and semi-detached houses had for a long time fallen short of demand and was now stepped up as other output diminished, which again probably helped to increase the number of vacant flats. Flat-dwellers left their rental flats in favour of houses, and landlords had difficulty in finding new tenants. Turnover was high, especially in utility-owned dwellings, which to a great extent were regarded by their inmates as a staging post on the way to something better.

It is fair to say that the aims defined immediately after the Second World War had been achieved by the mid-1970s. This coincided with the conclusion of the Million Homes Programme. After that the future of housing policy became uncertain (and still is in a manner of speaking), but initially it was assumed that housing policy would provide for basic social needs. The aims of equal costs for equivalent homes, the absence of wealth accumulation in housing, reasonable housing costs, a free choice of housing and user influence on housing conditions were defined in 1974. Interest in housing policy shifted partly from new production to the improvement and maintenance of older homes. A government commission (SOU 1984b) noted that large parts of the housing policy aims of the 1940s had been fulfilled but that housing standards varied considerably from one group of the population to another. The 1974 objectives, in other words, had not been achieved, and so housing policy should be reassessed and made to focus on the distribution, management and renewal of the existing housing stock. In addition, substandard housing environments should be improved and housing segregation counteracted.

The outcome of housing policy is summarized in Figure 2.1, which illustrates housing production. The impact of the Million Homes Programme between 1966

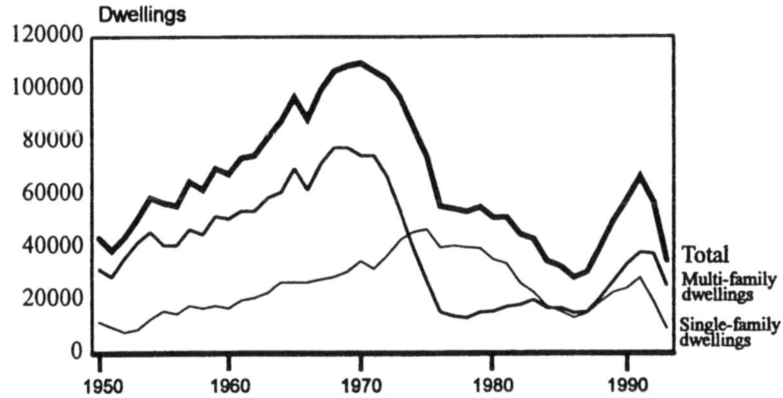

Figure 2.1 Dwelling production, 1950–93, by multi-family and single-family dwellings (Sources: SCB 1980, 1993a, 1994a).

and 1975 is clearly apparent, with a quarter of the existing housing stock dating from that period. One can also see that a fall in the output of flats was accompanied by rising production of detached and semi-detached houses. The same figure also shows that the post-1986 boom led to growth in output, since when, after a few years of recession, production has fallen off heavily.

State aid for housing production was amended in 1993 in such a way that developers perceived an advantage in making building starts before the end of the year. As a result, the number of units completed did not fall as steeply in 1992 as might have been expected in view of the current economic situation and the vacancy rate in the existing housing stock. The decline comes later – building projects commenced during 1993 involved only 11000 dwelling units, three quarters of them in apartment blocks (SCB 1994a).

To summarize, by international comparison Sweden has a high standard of housing and, in principle, all groups in society occupy good homes. Thus, 30% of the population in 1990 had more than one room per person at their disposal, in addition to a living room and a kitchen. Only 3% were living more than two persons to a room, still not including the kitchen and living room. Families with children have a lower spatial standard than all-adult households. Half the population live in apartment buildings, the rest in single-family dwellings. Two fifths own their homes, not quite one fifth are tenant-owners and two fifths rent their homes. Half the households in rented or tenant-owner flats are single-person households, as against only 15% living in detached or semi-detached houses (SCB 1993a).

2.4 Urban design and urban renewal policy

This section draws upon the description of urban design and renewal policy by Hall (1991) and Åström (1993), to which the reader is referred for a fuller treatment.

New urban areas

At the beginning of the 1930s, functionalism was launched as an alternative to the self-enclosing stone city. Sun, light and air became the slogans. Stockholm became a national path-breaker, and building took place on the fringes of the stone city and in completely new suburbs. Low-rise "lamella" blocks were built with green spaces in between, as well as detached houses. High-rise point or tower blocks, surrounded by greenery, were introduced later. Small commercial premises could be established in these new areas.

After the Second World War, the urban construction ideal was augmented with the idea of building up new areas out of smaller housing enclaves known as neighbourhood units *(grannskapsenheter)*, each with about 3000 residents; these could then be joined together, forming a subdistrict or ward of between 10000 and 15000 residents. The new districts were given reasonably adequate community centres as regards shops and public amenities. In the cities these districts could be joined together into even larger units, the ideal being for people to be able to live and work in one and the same district. Traffic segregation was essential for the avoidance of traffic accidents to children. The point blocks that had previously been so much in fashion were superseded by "lamella" blocks of between six and eight storeys.

Up until the Million Homes Programme of the 1960s, planning was dominated by functionalist principles, which meant separate zones for housing, peripheral industries and centrally located shops and offices. The neighbourhood idea was a living one, and the somewhat rigid planning structures of early functionalism had become more variable.

The crash housing production of the Million Homes Programme, however, gave large-scale, industrial building the upper hand. Trackbound cranes, large site depots and prefabricated elements called for a levelling out of the terrain. The buildings were standardized and complete areas were given identical volumes. Projects involving at least 1000 industrially constructed dwelling units qualified for extra-favourable credits. Building became a matter of technology and economics. Good housing came to be defined by national codes instead of the individual design of each project. Everything had to be done quickly and there was no scope for the creation of pleasant environments on a human scale. Minimal provision was made for outdoor spaces and amenities and communications often arrived long after people had moved into the new areas.

The homes produced were not always very attractive, and at the time the Million Homes Programme was drawing to a close, there was a glut of dwellings. Building capacity was diverted to meet the neglected demand for detached and terraced houses. Carpets of single-family housing were laid out around the localities, and often these were just as stereotyped as the multi-family housing areas constructed previously. Developers, both municipal and private, were as certain of finding a sale for these houses as they had been previously of finding tenants for the new flats. Standardized building production survived along with preparation of matching detailed plans, even though the buildings were now on a smaller scale. It was, incidentally, during this period that a reaction set in against these housing environments in the form of a "green wave" of migrants to the countryside.

It became increasingly obvious that the principles of both functionalism and industrialized building would have to be relaxed. Buildings and districts became too monotonous, and housing developments from the period between 1930 and 1980, when viewed from the air, often resemble circuit boards or computer chips. On the other hand it is perhaps not so surprising that, over and above the ethos of equality, planning was permeated by the belief that an optimum way could be found of combining the elements of dwellings, building volumes and districts or suburbs into complete towns and cities. The elements were to be united by means of hierarchical traffic networks based on scientific principles of accident prevention. Planning, like building, became over-susceptible to a functional, not to say semi-mathematical, mentality.

In the mid-1970s, and still more so in the 1980s, building development became more complex. Different types of building, tenure and activity could now be intermingled. The townscape and the architecture of buildings now became more elaborate, more exciting, gentler, often more intimate but sometimes chaotic. In some places, for example, there was a reversion to grid planning, based on enclosed blocks, whereas in other areas housing development became ecologically adapted. Urban development now demonstrated a diversity of expression that had previously not been very much in evidence.

Renewal of city centres and other ageing urban settlements

The heavy post-war emphasis on housing production led to a deliberate diversion of labour resources in that direction, and in fact building operations in city centres and other earlier settlement were sometimes prohibited, so that the resources would not be put into the restoration or development of existing environments. The discussion of urban renewal, once it began, focused mainly on the function of city centres and on the standard of existing, older housing. The urban environment as such was more grudgingly treated, especially as regards its traditional areas.

Some towns had already embarked on a renewal of their old centres in the 1950s and 1960s, even though this put something of a brake on the production of new housing. Gradually more and more towns and cities followed suit. Renewal took the form of total clearance – that is, the demolition of practically everything and its replacement with new buildings. Large units were aimed for, and detailed development plans often came to resemble those for greenfield sites. The heaviest pressure for renewal was exerted on the town centres, because of the large areas needed by retailers for self-service stores and also the preference of both private companies and public authorities for centrally located offices. In addition, growing volumes of traffic called for space for traffic and parking facilities. There was immense confidence in the future, and the aim was to plan and construct for the commercial and motorized city. All the department stores erected in central areas of towns, cities and other communities are emblematic of the period. Some of the urban centres that were renewed, however, became enclaves surrounded by traffic arteries, true to the principle of traffic segregation. This had the effect of cutting them off from the rest of the town or city.

Housing renewal in urban areas was looked on mainly as a matter of demolition and new construction. Conversion and improvement, as a rule, were conceivable only for buildings of outstanding historic or artistic interest. For a time, as we have seen, the renovation of old housing stock in many localities was actually halted on the grounds that the buildings were to be demolished in any case.

Defenders of tradition and of the preservation of old buildings did not gain a hearing in debate or practical policy-making until the beginning of the 1970s, when the great housing construction programme was drawing to a close. Periodic fierce criticism of renewal projects in progress was followed by a change of heart on the subject of urban renewal. In addition, the pressure from economically powerful interests was gradually relaxed as demand for new shops and premises became sated. The town centre renewals were not completed fully in accordance with the original plans; in the event, only a few blocks were affected. Large-scale projects were superseded by smaller-scale infill and renewal projects in pre-existing blocks. The word "clearance" (*sanering*) was superseded by "urban renewal" (*stadsförnyelse*). Parallel to this, action programmes were adopted for raising old housing to modern standards.

The government and the Riksdag also laid down new guidelines for housing renewal, to the effect that buildings were to be refurbished rather than demolished. In the mid-1980s, a nationwide programme was mounted and the rules of State housing finance were amended in support of it. Right from the beginning the aim was to upgrade old dwellings to modern standards. Social equality and integrated living were to be promoted and special attention was to be paid to enabling the elderly and disabled to carry on living in their accustomed surroundings. Traffic management was given priority, at the same time as efforts were made to bring about the renewal of residential courtyards. Compulsory clearance legislation was passed to deal with excessively negligent landlords.

As a result of the economic boom of the 1980s, public domination of building development was superseded by public–private co-operation. During this period, many older industrial estates and former transport operators' land reverted to housing and office use, with advanced external design. At the same time, the social ambitions of urban development were toned down to some extent and a spirit of entrepreneurship with elements of speculation emerged instead. That spirit was particularly noticeable in the non-housing sector.

2.5 Change of direction

Housing and land policies have come in for mounting criticism. Putting it very simply, two main lines are discernible. One of them is that the influence of those affected must be augmented at the expense of development interests. The other is that market forces should be allowed to govern developments more explicitly, this being the only way in which one can tell what housing consumers really want. Advocates of the first line of policy wish to create a more distinct, effective platform for bodies of opinion, whereas representatives of the second want to dismantle the public system of control and supervision. The system is also being accused of encouraging cumbersome and expensive planning processes. Some critics maintain that subsidized building escalates costs, while producing a glut of housing and also housing of an excessively high standard. From the viewpoint of government finance, the situation today is a difficult one.

A change of direction seems to have occurred at the beginning of the 1990s, and can be symbolized by the 1991 election, which led to the formation of a non-socialist administration. That administration abolished the Ministry of Housing and Physical Planning (*bostadsdepartementet*), transferring its responsibilities to six other ministries and thereby signalling that housing questions were not to have the same priority as before. The intention is for market forces to carry more weight, and this has triggered a debate as to whether municipality, developer or financier has and should have the last word on building development.

In the national policy context, work has begun on systematic changes, and proposals for the reform of planning, subsidies, building controls, the size of municipal organizations, and so on are to be expected within the near future. Several government commissions, for example, have been appointed to review land and housing policies in the broad sense. One prior assumption is that regulatory systems escalate production costs and that deregulations are therefore needed. A committee was appointed to make recommendations for derogating the planning system and the award of permits in connection with building, and at the same time to assess how a stronger influence could be exerted by the public at an earlier planning phase. In this connection the State is also investigating methods and organization for real-property subdivision. The introduction of freehold tenure for flats is being considered, and policy recommendations have been made in favour of moving towards market-level rents. The municipal housing companies are being transformed step by step into ordinary property companies, able to make profits and losses. The system of State credits support for housing loans has been altered and, as previously mentioned, environmental legislation is under review, although in this case with a view to achieving stricter control.

The State is also going through a thoroughgoing reorganization in general terms. Its property holdings are being incorporated, and national authorities nowadays rent their premises from State-owned companies, for the most part at going commercial rates. Other national government activities are also to be reconstituted on a company basis, and some sectors of State-owned enterprise are privatized. However, the government has indicated that heavy investment is to be made in the development of infrastructure during the next ten years, with roads and railways as the most conspicuous items. This expenditure is not really prompted primarily by actual needs and can be viewed partly as a labour market policy measure in the construction sector to counteract the rapid decline in housing and non-housing construction.

The municipalities are also creating property companies to take care of their holdings. The companies then rent the property back to the municipality. The same is happening with the county council properties.

The 1991 Local Government Act gave the municipalities a very free hand in matters of internal organization. This has led to an extensive reorganization of municipal activities, with the introduction of purchase and sale systems, the establishment of municipally owned companies, and privatization. Some of the housing utilities are selling off their properties, or attempting to do so. Some municipalities and, in certain cases, municipal companies have already transferred parts of their property holdings to renting companies on a sale–leaseback basis. This has been done, for example, with schools, sports halls, harbours, hotels and industrial buildings. The decline in the building sector has also led many municipalities to do away with parts of their planning capacity.

CHAPTER 3
The legal framework

3.1 The legal environment

Land use and changes of land use in Sweden are subject to extensive public regulations based on the following principles:

- Land use is governed by planning and the award of permits. In principle, a property owner is entitled to "enjoy" only his property within the bounds of current land use. That is, a change in the use made of the property requires permission from public authorities. This applies regardless of whether the land is used for housing, industry, agriculture and forestry, and so on.
- Purchase and sale, like enjoyment, of real property are circumscribed by regulations. Property boundaries may not be changed without official permission. In addition, several laws make it possible for national and local authorities, and also for private persons, to acquire land from a property owner by compulsory purchase, on certain conditions.
- There is legislation regulating the development, operation and financing of infrastructure, i.e. communal amenities necessary to the functioning of individual properties.

The Swedish planning system
Planning legislation in Sweden is made up of several enactments. The "paving" Act of this legislation is the Natural Resources Act (1987: 12), which provides guidance on the implementation of 12 specialized enactments: the Planning and Building Act (1987: 10), the Water Act (1983: 291), the Environment Protection Act (1969: 387), the Nature Conservancy Act (1964: 822), the Peat Deposits Act (1985: 620), the Road Act (1971: 948), the Electrical Installations (various provisions) Act (1902: 71), the Pipelines Act (1978: 160), the Civil Aviation Act (1957: 297), the Minerals Act (1991: 45), the Continental Shelf Act (1966: 314) and the Public Shipping Lanes and Harbours Act (1983: 293). All these enactments can have a bearing on urban development, but the most interesting of them in this connection is the 1987 Planning and Building Act.

The Natural Resources Act
The Natural Resources Act (*naturresurslagen*) lays down that land, water and the physical environment in general must be used in such a way as to promote long-term good management from ecological, social and economic viewpoints. Land and water areas must be used for the purpose for which they are best suited,

having regard to their character, their location and existing needs. Priority must be given to the kind of land used, which, from a public viewpoint, implies good management of resources.

The Act stipulates in general terms the types of land and water zone that *shall* be protected as far as possible. Zones defined as "national interests" shall be protected. The Act also contains what are termed "geographical guidelines" for certain specified coasts, mountains, rivers and lakes. These zones are "national interests" in which development undertakings and interference with the environment are permissible only if they are feasible without palpable damage being inflicted on the natural and cultural qualities of these zones. This provision does not impede the development of existing localities or local enterprise, but it does rule out large-scale industrial start-ups, for example.

The Act also requires government permission for certain major industrial facilities such as iron and steel complexes, metalworks, pulp and paper mills, refineries and heavy petrochemical industry, nuclear power stations and large facilities for combustion of fossil fuels. An application for permission to erect these facilities must include an environmental impact assessment (*miljökonsekvensbeskrivning* 1991: 650) of the facility concerned.

It is primarily the task of the municipality to determine whether to authorize the use of land and water zones for building development and to ensure that the proposed development is suitable, but the State is entitled to intervene in the event of "national interests" being threatened by municipal decisions. It is above all by means of the comprehensive plan that the municipality has to strike a balance between different interests and indicate how "national interests" are to be provided for (see below).

The planning system in the Planning and Building Act

Land use planning under the Planning and Building Act (*plan-och bygglagen*) is a municipal concern. Basically, the municipality alone decides where, when and how a plan is to be drawn up. That is, property owners cannot demand that the municipality draw up a plan for their land or appeal against the municipality's decision to refuse to draw up a plan. The government cannot make an order for the municipality to adopt, revise or cancel plans, except where necessitated by national interests or by interests involving several municipalities.

The compliance of buildings with the provisions of the Act is assessed by the municipality through planning and building permits. The system of planning and permits can be summarized as shown in Figure 3.1.

The Regional Plan (*regionplan*), which is not mandatory, can be used if necessary to co-ordinate the planning activities of several municipalities. These plans have not come to be extensively used in actual practice.

Every municipality must have a current Comprehensive Plan (*översiktsplan*) covering its entire area. This plan must indicate the basic features of land and water use and of urban (building) development, and must furnish guidance for more detailed planning and permission (Fig. 3.2). The comprehensive plan, however, is not binding. That is, it may be overruled by subsequent detailed

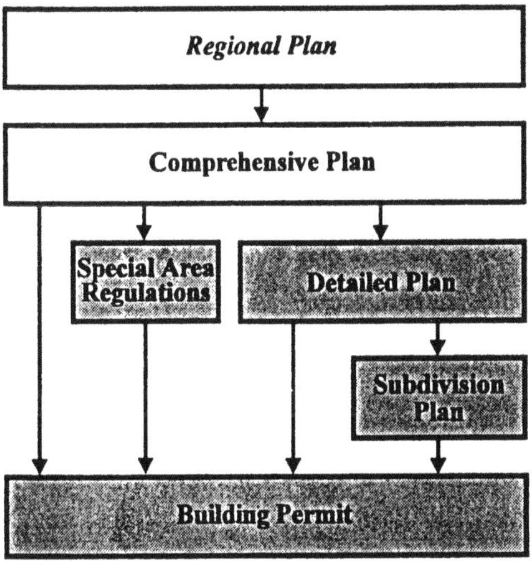

Figure 3.1 The system of planning and permits in the Planning and Building Act.

plans and building permits. If necessary, the plan can be made more detailed, or deepened (to take the sense of the Swedish term), for certain zones (*fördjupad översiktsplan* or *områdesplan*).

Special Area Regulations (*områdesbestämmelser*) are a regulatory procedure that can ensure, with binding effect, that the purpose of the comprehensive plan is achieved, which in turn means that building development contrary to its provisions *shall* be prohibited. These provisions are intended for use in limited zones for which a detailed plan is not to be drawn up. For example, essential use of land and water zones can be regulated in this way. Land use can be indicated for different types of development, e.g. recreational amenities, communications routes, restricted areas and safety zones.

A Detailed Plan (*detaljplan*) is mandatory for extensive changes of land use. This type of plan is prepared when a development has been initiated and has to be drawn up in the following cases: (a) for new, continuous building development that requires such communal facilities as roads and water and sewerage mains, (b) for a single new building that has considerable impact on its surroundings and/or is to be located in a zone where the pressure for development is high, and (c) for existing building development that is to be changed and preserved and needs to be regulated on an area basis. The amount of regulation needed concerning land use and building development naturally varies from one case to another, and the same is true of the content and appearance of individual detailed plans.

There are, however, certain mandatory particulars that the plan must always include:

- Public spaces (*allmänna platser*) such as streets, roads, squares and parks, must be delimited.
- Almost without exception, the municipality has mandatory power over public space, i.e. is responsible for the provision and maintenance of facilities.

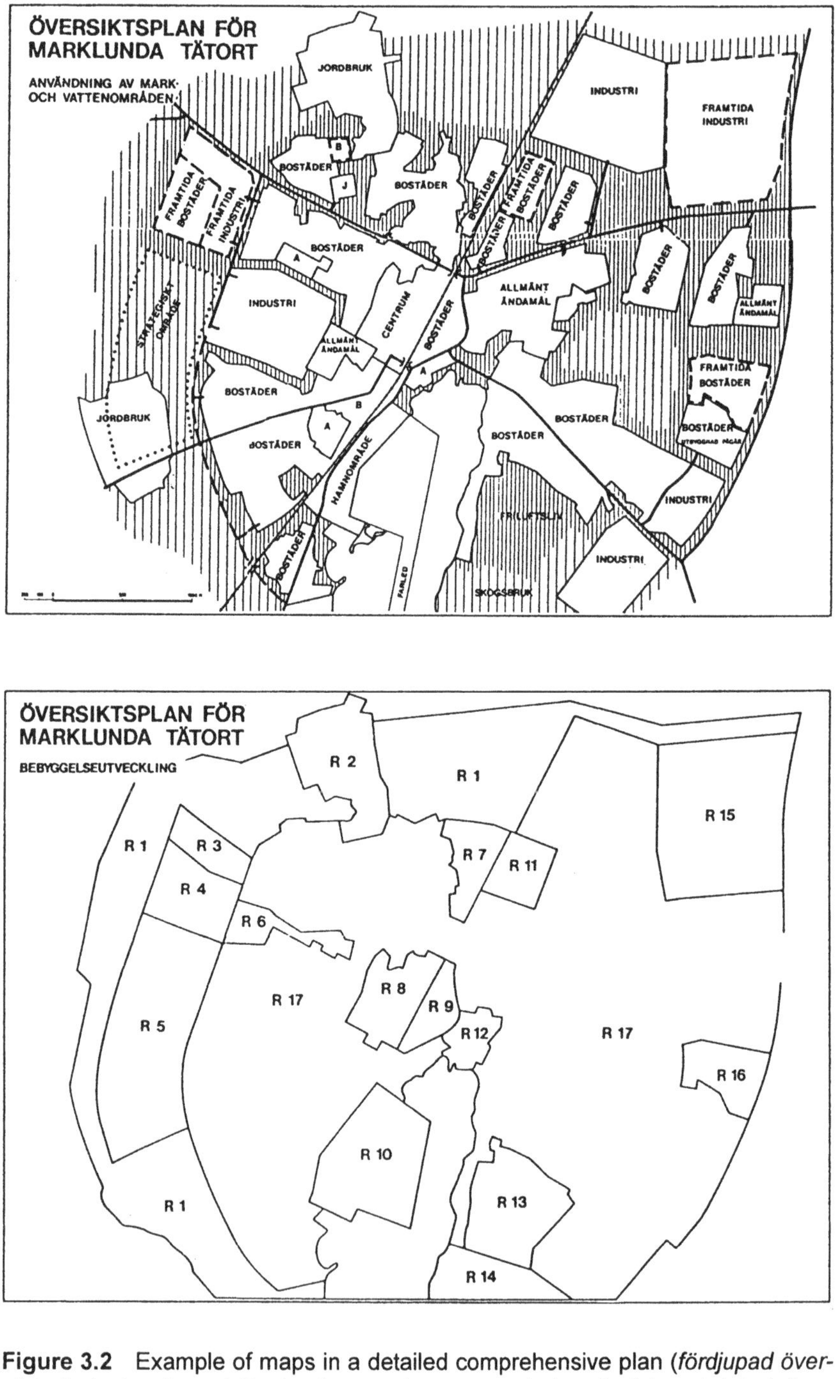

Figure 3.2 Example of maps in a detailed comprehensive plan (*fördjupad översiktsplan*), showing existing land use and recommendations for future detailed plans and building permissions in different areas (*Source:* Planverket 1987: 112)..

If the municipality is not so mandated, as can be the case with recreational development zones in the countryside, this has to be expressly stated in the plan.

- Building plots (*kvartersmark*) and the use to be made of this land for housing, offices, shops, industry, parking, community centre amenities, schools, and so on (more than 20 different categories of use can be specified) must be delimited.
- Detailed plans must have an implementation period (*genomförandetid*) of 5–15 years for the plan to be put into effect.
- In addition, where relevant, an account has to be given of water zones and their use as open water, technical installations in water (landing stages, bathing points, etc.) and harbour areas.

Over and above these mandatory particulars, a more precise definition is possible in the plan, for example of land use, the extent of settlement, its location, design and workmanship, and questions concerning land and implementation (see Fig. 3.3). For building land, for example, one can indicate the use to be made of individual buildings, the protection of existing vegetation, the location of vehicle exits, land to be preserved for parking facilities, pedestrian and bicycle traffic, and mains installations above and below ground. In order to control the extent, positioning and design of the building development, it is possible to delineate land that will not be built on, the height of buildings, the volume, colouring, shape, roof pitch, materials, mode of construction, noise suppression, energy management and the mix of differently sized dwelling units. Furthermore, a detailed plan may include provisions for the protection of existing buildings of historic, cultural, environmental or artistic interest. Preservation orders can be issued for existing buildings if they are needed for housing supply or are of cultural interest. For matters of planning implementation, land-parcelling principles such as plot size can, like communal facilities, be decided in the detailed plan. In many cases, though, it may be appropriate for these questions to be dealt with in a property regulation plan. The detailed plan provides scope for extensive regulation of land use and building development. The Planning and Building Act points out, however, that the plan must not be made more detailed than its purpose demands.

As stated above, the detailed plan is legally binding. This means that, when the plan acquires the force of law, rights and obligations are incurred especially by the municipality and property owners. A building permit (*bygglov*) application must, in principle, be granted if it accords with the plan. Furthermore, certain land in the plan can be acquired by compulsory purchase, for example for public space. The plan obliges the public space authority – usually the municipality – to construct and run the installations. Finally, decisions under certain other enactments must conform to a detailed plan. This applies, for example, to the Real Property Subdivision Act, the Joint Installations Act, the Environment Protection Act, the Nature Conservancy Act, the Water Act, the Road Act and the Civil Aviation Act.

The Subdivision, (i.e. property regulation or parcelling) Plan (*fastighetsplan*)

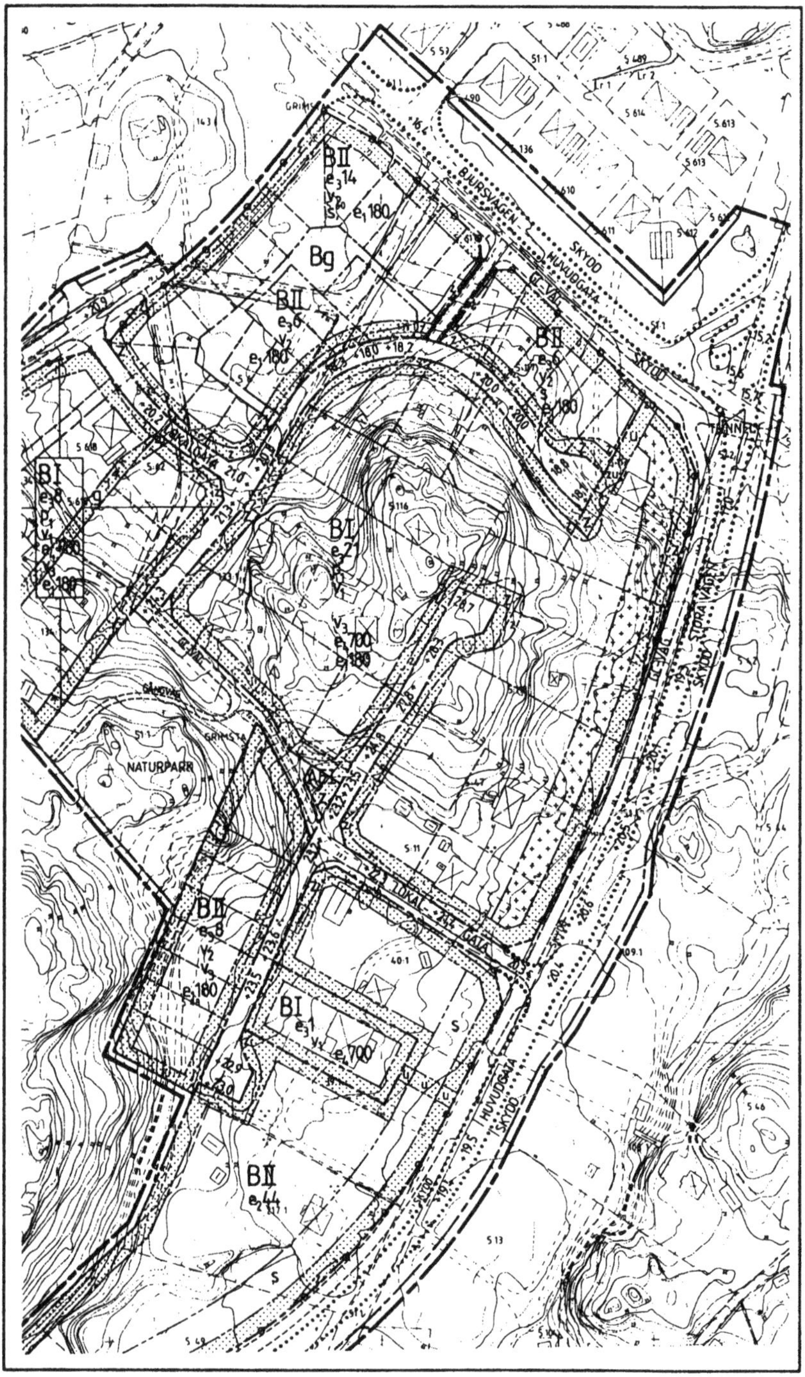

Figure 3.3 An example of the Detailed Plan: municipality of Upplands Väsby.

can be drawn up and adopted for zones with detailed plans, but it need not entirely conform to detailed planning provisions. There is a certain amount of scope for "minor deviations". The purpose of the subdivision plan is:

- To regulate property subdivision (property units, joint property units and easements). The plan is drawn up in order to achieve a co-ordination of future property subdivision – a co-ordination that would perhaps be impossible to achieve if each individual parcelling case were to be separately considered at different points in time.
- To regulate communal facilities shared by several property units. The plan can specify the facilities that are to be communal facilities, the property units that are to have a share in the facility, and the land used for the facility.
- To provide a means of assessing the suitability and location of utility easements (see Fig. 3.4).

The initiative in drawing up a subdivision plan usually comes from the municipality, but a property owner is also entitled to demand that a plan be drawn up (unless it is manifestly unnecessary). The documents normally belonging to a property subdivision plan are a map and provisions: a plan description, together with an account of existing conditions and the new property units/plots and joint property units to be formed under the plan, plus a list of the property owners.

The subdivision plan, like the detailed plan, has certain legal consequences. One prerequisite for the award of a building permit is that the property must conform to the subdivision plan. The subdivision plan also confers a right to the compulsory acquisition of certain land. For example, the person owning the greatest value of an intended property unit in the subdivision plan is entitled to acquire the remainder of the plot from his or her neighbour or neighbours.

In the overwhelming majority of cases, the detailed plan augments the building rights of the property owner. Sometimes, though, it imposes restrictions on the property owner, and in some such cases the question of compensation may arise if the property owner incurs "damage". This applies when:

- the detailed plan is amended or cancelled during the implementation period (if "exceptional detriment" is incurred by a property owner, they can request the municipality to acquire the property)
- the property owner, for reasons of traffic safety, is ordered to demolish a building, installation or device or to change a vehicle exit
- an existing building, which is at variance to the plan, is accidentally destroyed and building permission is not granted for a replacement
- an existing building has been made the subject of a preservation order in the plan, for example because it is needed for housing supply or is of cultural interest
- an existing building is demolished and no building permit is granted for its replacement with a similar building (a building permit has to be applied for, however, within five years of the building being demolished)
- current land use is considerably impeded by safety provisions for buildings of historic, cultural, environmental or artistic value.

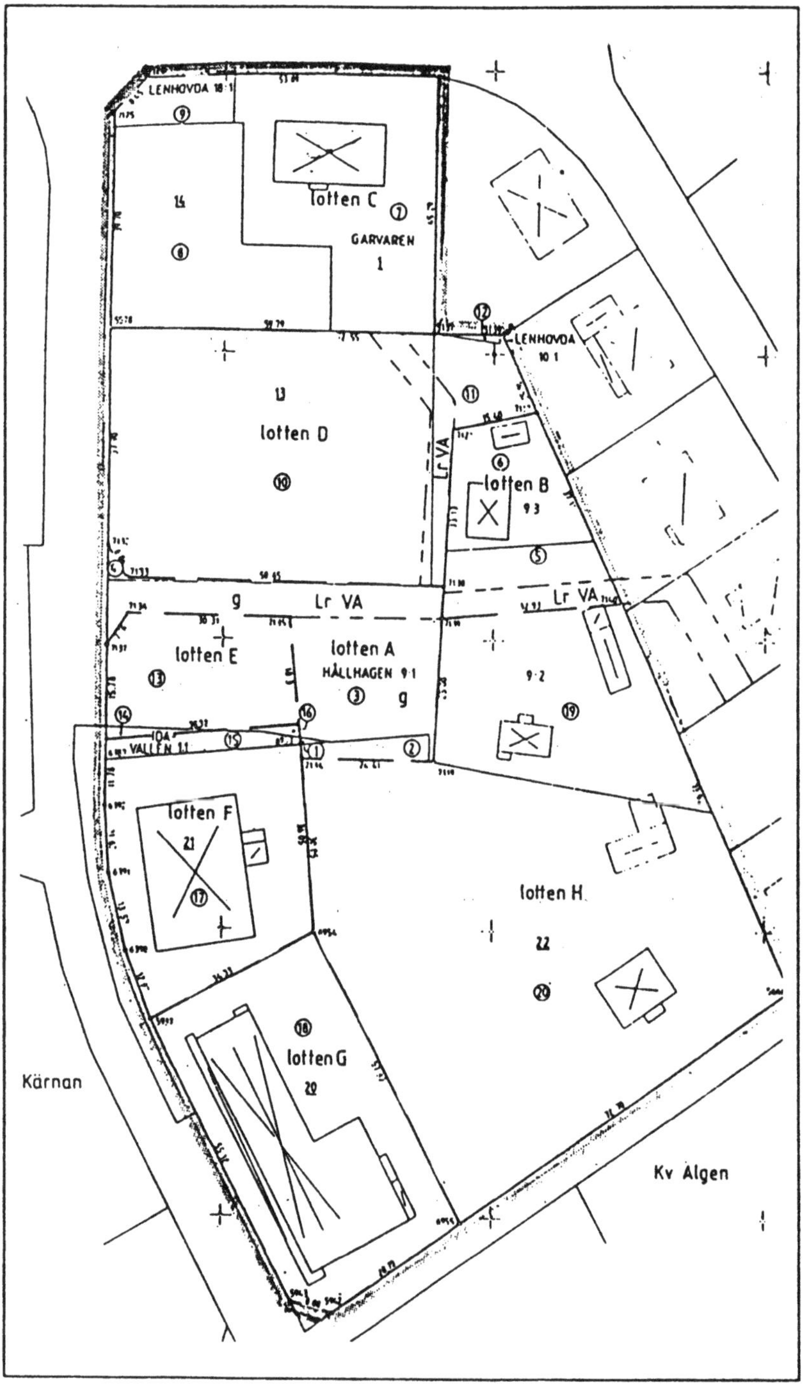

Figure 3.4 An example of the Subdivision Plan: municipality of Uppvidinge.

Development control by building permit procedure

A building permit (*bygglov*) is required for the erection of new buildings, for extensions, for the application of buildings to an essentially different purpose, for considerable changes in the layout of buildings, for alterations affecting load-bearing parts of buildings, and for the installation or essential alteration of fire-places, flues, ventilation, water supply, sewerage or lifts. In a zone to which a detailed plan applies, a building permit also has to be requested for changing the colour of façades, changing the roofing material, taking other measures that significantly affect the outward appearance of a building, erecting and significantly changing signage and lighting arrangements, and also erecting, extending or altering non-residential buildings for agriculture and forestry. Sometimes a building permit may be necessary for non-building structures too. This applies to the erection and alteration of walls, plank fencing, warehouses, depots and storage tanks.

Some minor measures are exempted from the general requirement of building permission. Single- and two-family dwellings, for example, can be painted a different colour, so long as the essential character of the building is left unchanged. It is also permissible to erect not more than two supplementary buildings, e.g. outhouses, garaging or other small buildings, if their total building area does not exceed $10\,m^2$ and they are at least 4.5 m from the property boundary.

A detailed plan provides opportunities for amending the general requirement of building permission. The plan may include provisions to the effect that building permits are not required for certain structures, e.g. for rudimentary weekend cottages or for the enlargement or alteration of industrial buildings. On the other hand, the plan may stipulate building permission for changing the colour of a single- or two-family dwelling or for maintenance work on buildings of outstanding value.

Building permit applications have to be made to the municipality and must include the drawings, descriptions, and so on necessary for processing an application. Building permits may not be granted for measures at variance with the detailed plan. On the other hand, a building permit must be granted for a measure in keeping with the plan, so long as that measure also meets certain requirements laid down in Chapter 3 of the Planning and Building Act, concerning the positioning, design and workmanship of buildings, and the property unit conforms to the subdivision plan (assuming there is one).

The stipulation that a building permit must not conflict with the detailed plan or the subdivision plan is not absolute. Minor deviations are allowed if they are compatible with the purpose of the plan. For example, a building can encroach by a metre or so on land that, according to the plan, may not be built on, and the height or area of the building may be exceeded for technical reasons, in order to achieve a better planning arrangement.

Apart from building permission, the detailed plan also requires permission for the complete or partial demolition of buildings subject to building permission (*rivningslov*). It is, however, possible to lay down through the detailed plan that no demolition permit is needed for certain buildings. On the other hand, through the detailed plan demolition permits can also be stipulated for buildings for which no building permission is required. As was explained earlier, outstandingly valuable building development can be protected by including a preservation order in the detailed plan. Within the detailed plan, permission is also required for excavation or landfill, which has the effect of considerably increasing the altitude of the ground (site improvement permission, *marklov*), unless the plan presupposes that the ground surface will be on this level. Through the detailed plan, the stipulation of site improvement permission can also be expanded so as to include tree felling and afforestation.

Environmental protection laws and other relevant acts

Several enactments are aimed at protecting the environment and have an effect on land development. Swedish environmental law can be divided into several subsectors:

- planning and land legislation
- protective legislation
- legislation on the extraction of natural resources, and
- legislation on the enlargement of certain installations.

Planning and land legislation The Natural Resources Act and the Planning and Building Act, the main substance of which has already been described, are the cornerstones of planning and land legislation. In addition there are several specialized enactments, some of which may come into play in the urban context.

The Municipal Energy Planning Act (*lagen om kommunal energiplanering* 1977: 439) requires the municipalities, in the course of their planning, to promote energy conservation and to achieve a safe and adequate energy supply. In its planning activities, the municipality is to investigate the feasibility of co-operating with other municipalities or some other interest in the energy sector, with a view to the joint resolution of questions relating to the conservation and supply of energy. Every municipality must have a current plan for the supply, distribution and use of energy within its boundaries.

The Planning and Building Act requires heritage conservation to be taken into account in municipal planning. But there is also the Heritage Conservation Act (*lagen om kulturminnen* 1988: 950), which affords protection for permanent archaeological remains such as graves, rune stones and rock carvings, and traces of dwellings, roads or footways and bridges. If a development enterprise affects an archaeological site, however, the county administrative board may grant permission for the removal of the remains. Such permission is granted on certain conditions, e.g. the investigation of the remains prior to their removal. Normally the cost of the investigation has to be borne by the developer. The same Act also makes provision concerning historic, listed buildings. A building that is of outstanding historic interest or forms part of an outstandingly valuable historic settlement area may be listed by the county administrative board as an historic building. The same goes for a park, garden or other structure of outstanding historic interest.

The Nature Conservancy Act (*naturvårdslagen* 1964: 822) includes several provisions for the protection and conservation of nature. Large, continuous areas can be preserved by designation as national parks or nature reserves. Natural objects and plant and animal species can be protected by means of protection orders. The Act lays down that the general public are entitled to move freely in the countryside, under the common right of access known as "Everyman's Right", so long as they do not cause damage. As a means of further securing public access to places for bathing and outdoor activities, protected shorelines exist beside the sea, lakes and watercourses (extending generally 100m from the shoreline, but extendable up to 300m). Within a protected shoreline area, the principle is that buildings or installations impeding public access to the shore may not be erected. Shoreline protection can be cancelled, however, in zones with special area regulations or detailed plans. Nature conservancy under the Nature Conservancy Act is both a national government and municipal concern. At county level it is the responsibility of the county administrative board.

Protective legislation This group of legislation includes several enactments with a greater or lesser bearing on land use and planning. They apply among other things to the handling of chemical products, e.g. the carriage of dangerous goods. Special rules apply to air pollution and to noise and other nuisances emanating from energy facilities and from road and air traffic. One group of laws deals with nuclear power production and radiation protection. That legislation will not be considered here in any further detail, but other vital enactments will be presented.

The Environment Protection Act (*miljöskyddslagen* 1969: 387) applies to the discharge of wastewater, solids or gas from land, a building or an installation, and also to more continuous disturbances from air pollution, noise, vibrations, light or suchlike. The National Franchise Board for Environment Protection (*koncessionsnämnden för miljöskydd*) can grant permission for environmentally hazardous activities on application being made by the person conducting or proposing to conduct them. The government has made certain types of environmentally hazardous activity absolutely conditional on the award of permits. About 70 such activities are listed in the Environment Protection Ordinance (*miljöskyddsförordningen* 1989: 364). That list also includes another 40 activities that are notifiable to municipal authorities.

The Health Protection Act (*hälsoskyddslagen* 1982: 1080) deals with measures to prevent and eliminate public health hazards, i.e. nuisances that may be harmful to human health and are not trivial or temporary. Among other things this Act applies to drinking and wastewater, pests, livestock protection and the handling of foodstuffs. Every municipality has to have a special authority to supervise its public health safeguards. Among other things the authority is required to draft proposals and to take part in planning with reference to matters of environment and health protection. In addition, it is empowered to issue injunctions and prohibitions for compliance with the Health Protection Act. Compliance with the Act at county level is supervised by the county administrative board.

The Refuse Collection Act (*renhållningslagen* 1979: 596) applies to the collection, storage, removal and final disposal of household refuse. This Act makes the municipality responsible for the satisfactory handling of domestic refuse. Municipal refuse management may be financed by means of charges paid by property owners in accordance with an annually adopted schedule.

Legislation on the extraction of natural resources The purpose of this legislation is to control the exploitation of the environment and at the same time to facilitate economically valuable exploitation. It can be divided into three subgroups. First there is legislation on water use, mainly comprising the Water Act (*vattenlagen* 1983: 291). Secondly there is legislation on accumulated natural resources, e.g. the Minerals Act (*minerallagen* 1991: 45), the Peat Deposits Act (*lagen om vissa torvfyndigheter* 1985: 620) and the Continental Shelf Act (*lagen om kontinentalsockeln* 1966: 314). And thirdly there is legislation on hunting and fishing, including the Game Act (*jaktlagen* 1987: 259).

Legislation on the enlargement of certain installations Enlargement of some installations is conditional on their being integrated with municipal planning. For all the types of installation referred to below, the rule is that permits may not be awarded contrary to special area regulations or detailed plans.

The Roads Act (*väglagen* 1971: 948) applies to public roads under State management. In principle this means the main road network outside localities. Within localities, streets and roads are normally a municipal responsibility, but certain major roads, e.g. through-roads, may come under State administration. In practice, public roads are managed by a national authority, the National Road Administration (*Vägverket*). Questions concerning the construction of public roads are dealt with by the National Road Administration in consultation with the county administrative board. For the construction of a road, a land acquisition plan (*arbetsplan*) has to be drawn up and accompanied by an environmental impact assessment. The assessment must contain an account of anticipated environmental effects and recommendations for necessary protective measures to prevent disturbance or other nuisances being caused by traffic (1987: 745). A land acquisition plan must be drawn up in consultation with the property owners affected and with public authorities and others whose interests are materially affected by the location and design of the road. Land needed for a public road may be coercively acquired by virtue of right of way. The compensation payable to the landowner is governed by the rules of the Expropriation Act.

The Electrical Installations (various provisions) Act (*ellagen* 1902: 71) applies to facilities for the generation, storage, transformation, transmission, distribution or use of electric current. High voltage power lines may not be constructed without a concession from the government (or an authority appointed by the government). A concession may refer to a power line following a mainly predetermined route (a line concession), or to a network of lines in a certain area (an area concession).

The Pipelines Act (*rörledningslagen* 1978: 160) applies to pipelines for district heat, crude oil, natural gas or any other liquid or gas to be used as fuel. The construction and use of such pipelines requires a concession (unless the installation is to be used solely for the requirements of individual households or exclusively within a port or an industrial perimeter). Concessions are awarded by the government.

The planning process for local plans

The compilation and adoption of local plans is closely regulated by the Planning and Building Act. The procedure is essentially the same, whatever the type of plan involved, but there are certain differences, for example, between the comprehensive plan and the detailed plan, and so a description of both processes is called for.

The comprehensive plan Local planning is a continuing process that has to be summarized every now and then, on a formalized basis, so that amendments to the comprehensive plan can be incorporated. The plan must be "current", but it is the municipality itself that decides whether or not this requirement is being met. The *travaux préparatoires* of the Planning and Building Act state, however, that the comprehensive plan should be reviewed in its entirety at least every five or six years. The procedure for drawing up the comprehensive plan is aimed at giving interests outside the municipal organization including residents, the State, interest organizations and pressure groups, neighbouring municipalities and insight into the planning process and an opportunity to influence it.

In this connection it is the *duty* of the municipality to consult the county administrative board and other municipalities affected whenever it is proposed to draw up or amend a comprehensive plan. The county administrative board must ensure that provision is made for "national interests", as well as generally securing the interests of the State. Others whose "vital" interests are affected by the plan "shall be given the opportunity" of consultation; this applies, for example, to large landowners, nature conservancy associations and the Federation of Swedish Farmers. The consultations have to be recorded in what is termed a "consultation report" (*samrådsredogörelse*).

- Before a comprehensive plan can be adopted, the draft version of it has to be exhibited for at least two months. The exhibition has to be announced in the local press at least one week in advance. During the exhibition, the draft plan must be accompanied by a plan description, which must present the planning issues, the purposes of the plan and the measures that the municipality is to take in order to give effect to it. A consultation report must also be included. During the exhibition, the county administrative board has to present a scrutiny report. That report must show:
- if the draft plan does not make provision for national interests
- if inter-municipal interests have not been co-ordinated, and
- if the development will be unsuitable, from the viewpoint of health and safety requirements.

Others wishing to tender viewpoints concerning the draft plan must do so *in writing* during the exhibition. After the exhibition period the municipality is required to issue a pronouncement summarizing the viewpoints received and indicating what effect they have had on the draft plan.

The comprehensive plan has to be adopted by the municipal council. Not being legally binding, it cannot be appealed against as far as its content is concerned, but residents can express dissatisfaction with the planning process by resorting to the local appeal procedure (*kommunalbesvär*).

The detailed plan Detailed plans can be drawn up in two ways: by normal or simplified planning procedures. The latter is possible if the draft plan is of limited importance, does not impinge on any public interests and is compatible with the comprehensive plan. Here we shall describe the sequence of the normal planning procedure. Schematically this can be illustrated as in Figure 3.5.

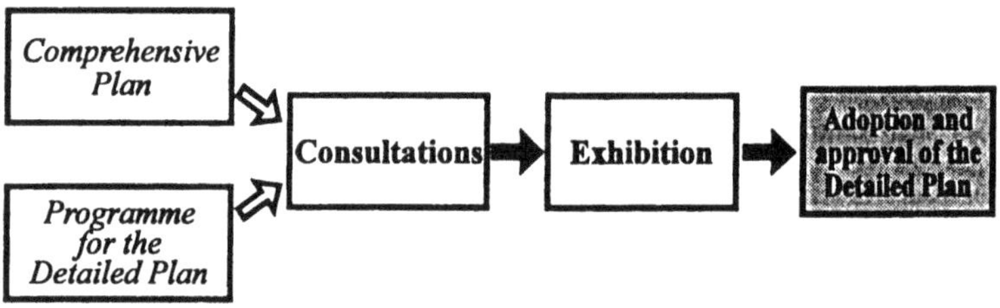

Figure 3.5 The basic procedure for compiling a detailed plan.

A detailed plan should be based on a programme containing a description of the present situation, the preconditions and purpose of the planning, the way in which planning work is to be conducted and a timetable for the planning process. Under certain conditions, a programme *has* to be drawn up. This applies if the use of land, buildings or installations implies "substantial environmental impact" and the detailed plan is drawn up for:

- industrial purposes
- new, continuous urban development
- ski lifts or cableways
- hotel complexes or holiday villages
- racing or testing circuits for motor vehicles.

In all these cases an environmental impact assessment must also be prepared in connection with the compilation of the detailed planning programme.

Consultation has two main functions in the planning process: first, knowledge and viewpoints can be supplied for the municipality concerning the planning area and the framing of the detailed plan, and, secondly, the municipality can supply information to those affected by the plan. The municipality *has* to consult with the county administrative board, the property registration authority and other municipalities that may be affected by the draft plan. The tasks of the county administrative board include ensuring that "national interests" are not counteracted (see the comprehensive plan, above). The municipality must provide an opportunity of consultation for property owners, tenant-owners, tenants and residents affected by the plan. The same applies to other authorities such as the roads administration, associations and private individuals whose vital interests are affected by the detailed plan. The consultation phase results in a consultation report (*samrådsredogörelse*), setting out the viewpoints emerging during the consultation procedure and indicating the stand that the municipality has taken on them. The main focus of the report should be on viewpoints that the municipality has not taken into account.

A draft detailed plan has to be exhibited for scrutiny for at least three weeks. The draft plan must be advertised on the municipal notice-board and in local newspapers at least one week before the exhibition period begins. In addition, notice of the announcement must be sent to property owners and others whose vital interests are affected by the draft plan. Before the exhibition takes place, a copy of the draft plan must be sent to the county administrative board and other municipalities possibly affected by the proposals.

The documents normally to be displayed at the exhibition include, for example:

- a planning programme if there is one
- a real-property list showing the owners of properties within the planning area and properties bordering on that area, and persons with special rights in a property, e.g. lessees and dominant tenants
- the consultation report
- a planning map with provisions, i.e. the main document of the detailed plan
- illustrations clearly showing how the planning area is meant to be used
- a plan description setting out the purpose and content of the plan and reasons for its actual design and provisions
- a developer's brief setting out economic preconditions and the technical, organizational and cadastral measures needed in order for the plan to be realized.

Anyone objecting to the plan must submit viewpoints *in writing* to the municipality during the exhibition period. Anyone wishing to appeal against the plan must be able to refer to written viewpoints that the municipality has not taken into account. After the exhibition the municipality has to collate the viewpoints received into a pronouncement. This pronouncement then has to be sent to persons whose objections to the draft plan have not been accommodated. If the draft plan is revised in essential respects on account of the exhibition, it has to be re-exhibited. If the amendment is of limited importance or of no interest to the general public, on the other hand, it can be processed by means of a simplified approval procedure, in which case the exhibition can be dispensed with.

When the municipality has adopted the plan, the resolution, together with instructions for appeal (*besvärshänvisning*), must be sent to (a) the county administrative board, (b) other municipalities in so far as they are affected by the plan, (c) property owners, tenant-owners, tenants, and so on who have tendered written viewpoints that have not been taken into account, i.e. those who are entitled to appeal against the planning decision, and (d) those who are entitled to appeal because the draft plan has been amended to their detriment *since* the exhibition. The content of a plan may be contested by appeal to the county administrative board, whose decision in turn can be contested by appeal to the government. The county administrative board for its part can review the municipal decision if national interests have not been provided for, land use in several municipalities has not been co-ordinated or the building development is unsuitable having regard to the health of residents or the risk of accidents. This decision by the county administrative board can also be contested by appeal to the government.

Private law relating to land, amendment of property boundaries

The Property Code (*jordabalken* 1971: 1209) represents the main body of Swedish legislation on real property. It contains rules on the conveyancing of property, on property as security for credit, on various forms of title to the property and of easements. The Real Property Subdivision Act (*fastighetsbildningslagen* 1970: 988) governs changes of various kinds in the subdivision of property and defines the conditions on which those changes can be made.

The Property Code The first section of the Property Code states: "Real estate is land. This is divided into properties . . ." Properties, then, are clearly defined areas of land. This means that they cannot be three-dimensionally stratified. For example, different dwelling units in a multi-storey building cannot be properties in their own right with separate owners. Nor is it possible for the area below ground to have one owner and the ground surface another.

A property includes a building, conduit, fencing and any other installation established in or above ground for permanent use (*allmänna tillbehör*). A building includes fixtures and other articles for permanent use, such as a lift, a water pipe, power sockets and a central heating boiler (*byggnadstillbehör*). An industrial property normally also includes machinery and other equipment used for the industrial activity (*industritillbehör*).

In principle, all land in Sweden is divided into properties and each property is registered by means of a unique property designation.

The right to buy and sell real property is unrestricted with certain exceptions (dealt with later in this section). There are, however, certain formal requirements without which a purchase is invalid:

- a *written* document of sale must be drawn up and signed by purchaser *and* vendor
- the document of sale must include an express provision whereby the vendor conveys the property to the purchaser, and finally
- the purchase price must be stated.

It must be noted that Swedish law does not countenance options in the form of a promise to sell or buy real property. The parties can, however, keep a purchase "in the air" by making the contract of sale include provisions whereby certain conditions have to be met before the purchase can be completed. The basic rule is that a sale cannot be kept in this state of suspension for more than two years. If the conditions are not met by then, the transaction is null and void.

The person acquiring a property must apply to have the acquisition registered within three months of the document of sale being drawn up, so as to obtain a title deed (*lagfart*). This application is made to the land registration authority, which checks to see that the formal requirements for the purchase are satisfied.

Purchase of real property need not refer to an entire property unit. It is also possible to acquire a share in a property or an area of land that does not constitute a property in its own right. In the first instance, the owners will acquire an intangible participation in the property and no distinguishable area on the ground. In the second instance, application for subdivision of the land area (or reallotment of the land area from the vendor's to the purchaser's property) must be made to the land registration authority within six months of the document of sale being drawn up. Failing this, or if the subdivision cannot be allowed, the purchase will be cancelled.

There are in general no restrictions on the right to acquire real property in Sweden, but there are one or two exceptions.

First, permission is needed for the acquisition of certain *agricultural properties*. This applies to properties in rural areas, if purchasers are not residents of the municipality where the property is situated, or if purchasers do not intend making the property their permanent residence. An acquisition permit must not be granted if the property is needed for the promotion of employment or settlement in the community. Furthermore, a permit is needed for the acquisition of properties in areas with highly fragmented ownership, if plans exist for rationalizing the ownership structure. Finally, all legal bodies have to apply for an acquisition permit before acquiring agricultural property.

Acquisition of *rental housing properties* is also subject to permission if the municipality takes the view that the acquisition should be examined. An acquisition permit is not to be granted, for example, if purchasers are unable to establish that they are in a position to manage the property. The purpose of this legislation is to ensure that rental housing properties will be acceptably managed.

Lastly, foreign persons and companies (i.e. non-EU) having no links with Sweden also have to apply for permission to acquire property for *recreational purposes*. This requirement applies to properties such as single-family dwellings and agricultural properties that may be used as holiday homes. An acquisition permit shall be granted if the property is required for a non-recreational purpose, e.g. if the purchaser intends using the property as a permanent home. An acquisition permit must also be granted in the event of acquisition for recreational purposes if the acquisition does not conflict with a "vital public interest" (the *travaux préparatoires* of the Act state that the question of refusing a permit on these grounds can be expected to arise only in rare cases). On the other hand, an acquisition permit shall be refused if the property is located in an area where recreational properties are so much in demand that, for this reason, there is a risk of property values escalating.

Real property can be pledged as security for credit. Basically this system operates as follows. The first stage is for a mortgage deed (*pantbrev*) to be issued at the property owner's request. This mortgage deed, which belongs to the property owner, shows the date of its issue and the amount to which it refers. These particulars are entered in the land register as soon as the mortgage deed is drawn up. As a second stage, the property owner can use the mortgage deed as security for credits, by surrendering it to the creditor. The more detailed credit conditions - interest, payment period, and so on - are governed by a special debt instrument drawn up by the property owner and the credit grantor. In many cases, several mortgage deeds have been drawn up for the same property. If so, the order of priority between different mortgage deeds will be based on the chronological order in which they were drawn up, so that an earlier deed will have priority over a later one.

Property owners can assign a beneficial interest in their property to another person, a company, and so on by contract. This beneficial interest can take the form of site leasehold (*tomträtt*), leasehold (*arrende*) or tenancy (*hyra*).

Site leasehold means that the municipality (or the State) convey land to the site lessee for an indefinite period, in return for an annual ground rent. This ground rent must remain unaltered for a certain period of time, which must not be less than ten years. The purpose of site leasehold is to enable the municipality to benefit from a rise in the value of the land. Otherwise, site leasehold is very similar to freehold. Site lessees are entitled, for example, to sell their title. They can also mortgage it.

Leasehold means the conveyance of land for a certain fixed period in return for a consideration. The Property Code recognizes four types of leasehold. Agricultural leasehold (*jordbruksarrende*) means the leasing of land, and in some cases of buildings, for agricultural purposes. (Understandably, this type of leasehold is of minor importance in the urban context.) Residential leasehold (*bostadsarrende*) means the conveyance of land for dwellings. This is employed primarily for recreational development. Non-residential leasehold (*anläggningsarrende*) refers to the leasing of land for commercial activities, e.g. in the form of a factory, warehouse, petrol station or kiosk. Lastly there is property leasehold (*lägenhetsarrende*), a catch-all term for leaseholds not coming under the other three categories. Statutory requirements concerning the form of the leasehold agreement, its duration, and so on vary from one type of leasehold to another, the most extensive regulations being those for agricultural and residential leaseholds, whereas property leaseholds are, in principle, entirely a matter for negotiation.

The term "rent" is used in the Property Code to refer to the conveyance of a building, or part of a building, for use in return for a consideration. Rental tenancy rules govern the form of the tenancy agreement, the term of the tenancy and cancellation of the tenancy agreement, security of tenure for the tenant, rent levels, and so on. Both residential and commercial premises can be rented. In view of the very great social importance of housing rents, with rented homes constituting about 40% of Sweden's housing stock, the statutory provisions for the protection of housing tenants are a great deal more comprehensive than those for non-residential tenants.

A distinction has to be made between rental tenure and tenant-ownership (*bostadsrätt*). Tenant-ownership is a beneficial interest in a building or part of a building and resembles tenancy in that a consideration is involved, but differs from it in several other vital respects. A tenant-owner title can be conferred only by a tenant-owner's association, which is a form of co-operative housing enterprise consisting of the members or tenant-owners in the association. Legally speaking, the association owns all the dwellings, whereas each member has a beneficial interest in - or tenant-owner title to - a specified dwelling. However, tenant-owners can sell their title. Tenant-ownership is a very common form of residential tenure, applying to some 15% of Sweden's housing stock.

An easement (*servitut*) entitles one property, for an indefinite period, to use another property for a certain purpose, e.g. for a road or utility. A written, civil easement contract (*avtalsservitut*) can be entered in the land register so as to protect the easement when the respective property changes hands. A distinction has to be made between contracted and official easements. The latter are formed by official decisions, e.g. through executive proceedings conducted by the Property Subdivision Authority (*fastighetsbildningsmyndigheten, FBM*), after which they cannot be cancelled without renewed official assessment; see below.

The Real Property Subdivision Act Changes in property boundaries come under the Real Property Subdivision Act (*fastighetsbildningslagen* 1970: 988) and can be made only by specially competent surveyors from national, and in certain cases municipal, Property Subdivision Authorities. Changes can mean the formation of new property units or the recasting of old ones. A new property unit can be created in several ways:

- an area can be detached from an existing property unit (parcelling, *avstyckning*)
- a property unit owned jointly by several persons, each having a certain share in it, can be distributed in such a way that each part-owner's share becomes a separate property unit (partition, *klyvning*)
- the owner of several property units can amalgamate them into a single unit (merging of titles, *sammanläggning*).

Reallotment of property units (*fastighetsreglering*) can be brought about by transferring an area of land from one property unit to another, and forming what are known as official easements. Subject to certain conditions, reallotment is possible against a property owner's wishes.

When changes are made to the subdivision of property, the new property units must be permanently suitable for their purpose and must conform to any pre-existing plans. Therefore, the surveyor has to consult the municipality in matters of property subdivision. Whenever a change is made to the property unit, a map is drawn up showing the unit's boundaries. Buildings are often shown as well. The corners of the property unit are normally marked on the ground, e.g. with metal tubes. The map also includes a description of the changes made to the property subdivision.

Instruments for the implementation of plans

Changes of land use mean, as a rule, that both owners and property units have to be adapted to new conditions. In other words, land has to be conveyed and, in addition, communal facilities such as streets, green areas, water and sewerage mains have to be constructed, managed and financed. In this section we will be describing statutory rules on compulsory purchase of land and on the development of local infrastructure. We will also be concerned with development agreements between the municipality and a property owner/developer, and with the principles determining who is to benefit from the appreciation resulting from land development. Lastly we will consider various provisions aimed at preventing activities that could otherwise impede current planning.

Land acquisition The moment when the detailed plan becomes legally binding is a vital stage in the development process. This being so, it may be appropriate to distinguish between land acquisitions made respectively before and after the detailed plan. These two types of acquisition can be characterized as (a) strategic land acquisitions before the detailed plan and (b) land acquisitions for the purpose of planning implementation after the detailed plan. Sweden has several laws containing rules on compulsory acquisition of land. Where strategic acquisitions before the detailed plan are concerned, it is mainly the Expropriation Act and the Pre-Emption Act that apply. Rules on compulsory acquisition of land for direct purposes of planning implementation are contained in the Planning and Building Act, the Property Subdivision Act, the Joint Installations Act, the Utility Easements Act and the Joint Land Development Act.

Under the Expropriation Act (*expropriationslagen* 1972: 719) it is above all the State and the municipalities that are empowered to acquire land by compulsory purchase for a variety of purposes, e.g. communications, entrepreneurial activity, restricted areas and safety zones, defence, nature conservancy and outdoor recreation. The municipality can also expropriate property units that will be needed for "urban settlement" later on, e.g. land for housing, offices or factories and supplementary communal amenities such as roads and green spaces. The municipality may expropriate land for urban settlement only if it is able to establish that the land is needed for the intended purpose. This can be done, for example, with reference to the comprehensive plan. In order to protect private interests, i.e. the interests of property owners, the power of expropriation is subject to a general restriction. Expropriation is not permitted if "the purpose ought properly to be provided for in some other way or if the inconveniences entailed by expropriation from a public and private viewpoint outweigh the advantages that can be gained through it".

Permission for expropriation is normally given by the government. In matters of "minor importance" it can be awarded by the county administrative board. It should be pointed out that expropriation is relatively unusual. For example, the total number of expropriation applications in 1987 was just over 100 (Bjerkén 1990: 121).

The main rule concerning compensation to people forced to surrender their property is that it must equal the market value of the property. If only part of the property is expropriated, the compensation must match the market value reduction that the expropriation entails. But this compensation does not give full coverage for the economic damage incurred by the property owner; compensation must also be paid for what is termed "other damage". There are, however, two important exceptions to the market value principle. First, the positive or negative impact, if any, of the expropriation enterprise itself on the property must be disregarded. Secondly, no compensation need be paid for expectation values, i.e. values based on expectations, arising less than ten years before the expropriation, of a change in the permissible use of the property.

Through pre-emption the municipality is entitled in certain circumstances to "take the purchaser's place" when the property changes hands. In this way the municipality is entitled to take over the property on the terms and at the price agreed on by the seller and purchaser. The Pre-emption Act (*förköpslagen* 1967: 868) empowers the municipality to exercise its right of pre-emption if a property:

- is needed for future urban settlement
- is in need of improvement or alteration to provide necessary housing
- is needed for sporting and outdoor recreational activities
- has a building that should be preserved for historic or environmental reasons
- has a building that needs to be used as a permanent home, if the property is situated in an area where second homes are very much in demand.

Properties of less than 3000 m^2 with single-family dwellings at the same time constituting permanent or recreational homes for not more than two families are, however, exempted from pre-emption. Nor is pre-emption possible with the State as vendor/purchaser or if the sale is between husband and wife. There is also a general prohibition against exercising the right of pre-emption in a way that is "oppressive" (for example, if the vendor wanted to do the purchaser a favour by selling the property at a bargain price, as opposed to the market price). If the vendor or purchaser objects to pre-emption, the municipality has to apply to the government for a pre-emption permit. Relatively few pre-emption cases - something like 20 per annum - were referred to the government during the 1980s.

Compulsory acquisition under the Planning and Building Act is governed by Chapter 6 of the Act. Those provisions entitle the municipality to acquire land reserved in the detailed plan for public spaces, for which the municipality is responsible and as such responsible for construction and maintenance. The compensation payable to the property owner must be based on the market value of the land, given its permissible use immediately before the detailed plan. Land for public spaces that will *not* come under mandatorship (legal competence or duty) of the municipality (e.g. roads and green spaces in a second-home development area) cannot be requisitioned under the Joint Installations Act (*anläggningslagen* 1973: 1149). That Act deals with what are termed communal facilities, i.e. facilities common to two or more property units (see below). Through this legislation, the property owners can requisition the land or space needed for the facility. The amount of compensation is governed by the same rules as apply concerning public spaces under municipal mandatorship. The Planning and Building Act empowers the municipality to acquire a development site that is not reserved in the detailed plan for "private building". This can mean public buildings, schools, day nurseries, sports and recreational facilities, railway and other traffic installations, areas for harbour, energy production and water and sewerage amenities, and so on. One precondition for compulsory acquisition of the land by the municipality, however, is the inability of the landowner to prepare land for the purpose intended. Compensation here is payable at the market value, depending on the use that would probably have been made of the property if the land had *not* been designated a public development site (although compensation must never fall short of the value of the land as currently used).

Communal facilities on a development site come under the Joint Installation Act and can be of various kinds, e.g. access roads, utilities, parking facilities, garaging and neighbourhood amenities. Compensation must be gauged according to what the value of the property would probably have been if it had not been designated as a communal facility. Here again, the minimum rule is that the property owner must always receive compensation equalling the value of current land use.

When a property subdivision plan is drawn up for an area, the new subdivision may deviate from the old one – that is, a new plot, as shown by the plan, may have to be composed of parts of two or more property units. By virtue of the subdivision plan and the Real Property Subdivision Act, the owner of the most valuable property unit fraction within the new plot is entitled to acquire the remaining parts by compulsory purchase. Payment in this case must be determined according to the land value of the new property unit following the adoption of the new plan. Furthermore, the basic assumption is that every square metre of the new property unit is of equal value. That is, the compensation payable will equal the area of the land conveyed multiplied by the value of the property unit per square metre. In a detailed plan, development land can be reserved for public footpaths and cycle paths and for public motor traffic through a tunnel or on a bridge. Space can then be requisitioned by means of an easement under the Real Property Subdivision Act. Alternatively, space can be provided by means of an easement or usufruct under the Expropriation Act, in which case compensation will equal the difference between the value of the property, once the land has been requisitioned, and what it would probably have been worth if it had not been reserved for traffic purposes.

Anyone needing access to land for utilities, both overhead and below ground, can obtain a title to land in the form of a utility easement (*ledningsrätt*) under the Utility Easements Act (1973: 1144). This applies, for example, concerning:

- a water or sewerage main included in a public water and sewerage facility
- a district heating main for public requirements
- a high-voltage power line for which a concession is required
- a public telephone line or any other public low-voltage line.

Conveyancing of utility easements is handled by a Property Subdivision

Authority and compensation must equal the difference between the value of the property after land has been provided for the utility and what the property would probably have been worth otherwise.

A utility easement can also be created for a property owner's private water and main sewers across another property so long as the utility entails only "a minor intrusion" in relation to the benefit. As an alternative method, an easement can be created under the Real Property Subdivision Act, in which case compensation is governed by the principles already described.

In certain cases a property owner may not have applied for building permission for the property unit during the implementation period of the detailed plan. In other words, the property has not been developed in accordance with the plan during the implementation period. If so, the municipality is entitled to compulsorily acquire the whole or parts of properties that have not been developed "essentially in agreement" with the detailed plan. The same power applies to parts of two or more property units that are to make up a new plot under the property subdivision plan. Compensation for the land must equal the market value of the property unit after the expiry of the implementation period.

The Joint Land Development Act (*lagen om exploateringssamverkan* 1987: 11) is a new body of legislation that came into force simultaneously with the 1987 Planning and Building Act. Joint development is mainly intended as an alternative to "conventional" implementation of detailed plans in areas with several property owners. This can include development of greenfield sites and the renewal of previously developed areas. The Joint Land Development Act makes it possible for property owners to collaborate in developing an area. Through the joint "land development unit" (*exploateringssamfällighet*), as it is called, the property owners can, for example, lay out streets, urban open spaces (green spaces) and water and sewerage mains, which are subsequently transferred to the municipality. The rules concerning land acquisition and compensation, however, are more interesting for present purposes. In certain development situations it may be desirable to draw up a detailed plan that requires a more thoroughgoing revision of the existing property boundaries. The rules of the Joint Land Development Act make it possible for restructuring of this kind to be carried out on an "equitable" basis. Each property is allotted a participation unit (*andelstal*) that, in principle, must correspond to the property unit's area. The participation unit has to be used for distributing the expenses and income of the development, i.e. the development profit, between the property units. It must also form the basis for the allocation of building land to which the property units are entitled. In this way, through reallotment or land transfer, property owners can be assured of land for individual building regardless of whether their original land is to be used as a development site, for a street, for green spaces and so on.

Development of local infrastructure A detailed plan, of course, includes several types of facility that are common to the property units - streets, green spaces, water and sewerage mains, district heating pipelines, electricity supply and telecommunications. There are many laws regulating the construction, operation, financing, and so on of these facilities. Normally the municipality must be the mandator of public spaces in a detailed plan - such as streets, play spaces, green spaces (Ch. 6, Sections 31–39, the Planning and Building Act). It is then the duty of the municipality to construct public spaces parallel to the completion of the public development and in accordance with "local custom". To cover the cost of public spaces, the municipality is entitled to levy charges on the property owners. These charges must not exceed the construction cost of the facilities. Maintenance costs, on the other hand, have to be funded out of local taxation revenue. Moreover, charges can be made only for "facilities serving the area". That is, the facilities must exclusively benefit the properties that are to pay the charges. This means, for instance, that the cost of streets and parks serving larger areas, for example through roads, service roads and high streets, has to be met out of taxation revenue. Street expenditures are recouped by the municipality compiling and adopting a "street construction cost report" for each individual detailed plan. Charges become payable by the property owner when the public spaces are completed. Implementation of the rules on the recouping of street construction costs varies a good deal from one municipality to another (Larsson 1988). Many municipalities do not charge property owners anything at all for these costs. Among those that do, the cost coverage rate varies between 40% and 100%. It is above all municipalities in expanding regions that do not subsidize property owners.

Public water and sewerage facilities come under the Water and Sewerage Act (*lagen om allmänna vatten- och avloppsanläggningar* 1970: 244). Water supply and sewerage are mostly managed by the municipality, but there are instances of public water and sewerage systems being operated by regional inter-municipal associations or even private mandators. The obligation of the municipality to construct public water and sewerage facilities, unlike its duty to build streets and other public spaces, is not formally geared to the detailed plan. The municipality must install public water and sewerage mains if on the one hand they are needed for reasons of public health (e.g. if the existing wastewater management system presents a public health hazard) and, on the other, water supply and sewerage for existing or impending settlement must be arranged in a wider context, which usually means that the installation is to serve at least 20–30 properties (Bouvin & Hedman 1972: 78). A municipality defaulting on its duty to enlarge a water and sewerage system can be ordered to do so by the county administrative board. The geographical areas within which the municipality operates water and sewerage facilities are termed water and sewerage management areas. The overwhelming majority of urban settlement comes within management areas of this kind, and these facilities are utilized by almost 90% of households (Larsson 1988: 18).

The municipality is entitled to levy set charges on property owners. The total charges may not exceed the cost of the water and sewerage installations to the municipality. Cost coverage for all municipalities in Sweden averages about 80%, but the larger municipalities, as a rule, have full cost coverage. The charges consist of a non-recurrent connection charge and an annual user charge.

The connection charge, which may not exceed the cost to the municipality of constructing the water and sewerage installation, frequently covers the cost of the local network. The cost of main distribution networks, purification plants, and so on and running costs, on the other hand, are financed out of the user charge. The connection charge is often based on the following parameters: (a) a basic charge that is the same for all properties, (b) a charge for the dwelling unit and (c) a charge per square metre of plot area. The user charge is generally based on (a) an annual basic charge per property unit and (b) a variable charge relating to the quantity of water supplied/consumed. It should be emphasized that the water and sewerage charge has to be computed with reference to the costs of the entire management area, and not just those occurring in the development area concerned. This means that connection charges for a new development can either exceed or fall short of the cost of the local network for the area. In this respect, then, water and sewerage charges differ from charges for street construction costs, which are based on the facilities constructed in the development area. The structure of the water and sewerage charge is determined by the municipality every year. The municipality is entitled to levy a connection charge when the water and sewerage installation is completed and the property owner has been notified that the property can be connected to it.

"District heating" in Sweden is a collective term for large-scale municipal production and distribution of heat in the form of hot water or electrical energy. District heating operations are managed by municipal utilities or companies. Larsson (1988) states that upwards of 150 municipalities have district heating and that, all in all, about half of the multi-family dwellings and non-residential properties in Sweden are connected. Less than 10% of single-family dwellings, however, are served in this way. Altogether, then, district heating provides roughly one third of the national heat supply (Svedinger 1989). The number of detached houses connected to district heating networks is relatively small, but on the other hand three out of every four new homes in multi-family dwellings are connected to district heating networks. The rules for municipal district heating systems are laid down in the Public Heating Systems Act (*lagen om allmänna värmesystem* 1981: 1354), which is constructed on essentially the same lines as the Water and Sewerage Act. Despite the important part played by municipal district heating in energy supply, not one single district heating plant comes, formally speaking, within the purview of the legislation. Use of these facilities is governed instead by voluntary agreements between the municipality and property owners. This is probably attributable above all to the fact that a declaration of public status involves the municipality in legal obligations. For example, the municipality could then be forced to enlarge a facility within the management area and also to compensate a property owner for individual heating installations that are made superfluous when the property is connected to the district heating network.

Power and telecommunication lines come under the Electrical Installations (various provisions) Act (*ellagen* 1902: 71) and the Telecommunications Ordinance (*teleförordningen* 1985: 765). The local power supply system is connected to the national grid. Some municipalities have their own power production facilities, but the majority are dependent entirely on the national grid, which is fed mainly by hydropower and nuclear power. Power is supplied from the national power network by Vattenfall (a State-owned undertaking) or private companies to power distribution companies, which in turn supply power to the individual property units. About half of all power distribution undertakings are municipal utilities or companies. Power supply is financed entirely by means of charges to users. These generally take the form of a connection charge, an annual fixed charge and a variable charge based on actual consumption (Svedinger 1989: 73).

The telecommunications system comprises a national network, a local network and the subscriber's network for the individual property unit. State-owned Swedish Telecom (*Telia*) formerly had a monopoly of telecommunications services, but this has been gradually abolished. The public telecommunications network will continue to be administered by Swedish Telecom, but other undertakings will also have access to it. Telecommunication lines are also financed entirely by means of charges in the form of a connection charge, an annual subscriber charge and a unit charge. The structure of telecommunication charges is determined by the government, whereas other services using the telecommunications network are priced commercially.

Communal facilities (*gemensamhetsanläggning*) are facilities common to several property units and managed by the property owners themselves. Facilities of this kind come under the Joint Installations Act and almost 40000 communal facilities have been established (CFD 1993).

Public spaces in a detailed plan, which do not come under the municipal mandate, have to be constituted as communal facilities as roads, green spaces, and so on. A communal facility can also comprise roads, footpaths, and play spaces on a developed site used by several property units. It can take the form of a water and sewerage system, power and telecommunication cables, heating installations, and so on, regardless of whether the facilities are located in a public space or on a developed site. Altogether there are nearly 80 different types of amenity included in a communal facility (Kalbro & Larsson 1983). The commonest amenities, in addition to those mentioned already, are parking, garaging, outdoor lighting, storage premises, neighbourhood facilities and laundries.

The establishment of communal facilities is handled by the Property Subdivision Authority. In order for a communal facility to be established, certain conditions have to be met:

- a communal facility must be of "essential importance" to the properties
- a communal facility can be established only if the "benefits" exceed the "cost and inconvenience" that the facility entails
- a communal facility may not be established if there is widespread opposition to it among the property owners for significant reasons.

In addition, a communal facility may not be set up at variance with a detailed plan or property subdivision plan (except for minor deviations). The cost of a communal facility must be allocated according to the participation units (*andelstal*) allotted to the property units. First, the construction cost of the facility must be distributed according to the "benefit" that the properties derive from it. The term "benefit" as used here refers to the appreciation in the value of the property resulting from the facility. Secondly, operating and maintenance costs must be distributed in relation to the "use" of the communal facility by the different properties. The establishment of communal facilities is confirmed in a construction order (*anläggningsbeslut*), which among other things states the purpose and location of the facility, the properties to participate in it and their various participation indexes. The construction and operation of the communal facility must comply with the construction order and are otherwise governed by the Joint Property Management Act (*lagen om förvaltning av samfälligheter* 1973: 1150). This provides two forms of management. In the case of co-owner management (*delägarförvaltning*), the property owners must agree on all measures to be taken, which in turn presupposes a relatively small number of owners. Larger communal facilities, however, are run on a condominium (*föreningsförvaltning*) basis, in which case basic management policy will be decided by a general meeting, and implementation of these decisions, together with the day-to-day running of the facilities, will be entrusted to an executive committee.

As was shown earlier, the Joint Land Development Act (ESL) provides a means of altering property boundaries in areas with several property owners. Another important purpose of ESL is to enable property owners to join together in constructing streets, green spaces, water and sewerage installations, and so on. The possibility of property owners jointly developing facilities is not specific to ESL. As has already been made clear, the Joint Installations Act can also be used for this purpose. The scope of this Act, however, is circumscribed in that the property owners may construct only communal facilities - that is, facilities that are to be operated and maintained by the property owners themselves. ESL opens up the possibility of the property owners also constructing facilities that will later be managed by the municipality, e.g. streets and other public places, municipal water and sewerage installations. Through the joint development unit, the property owners can act as *one party* in relation to the municipality, and so the joint development unit and the municipality can enter into an implementation agreement (see below), which among other things will make provision for the enlargement of municipal engineering installations.

Development agreements Previous sections have already shown that many rights and obligations of developers, property owners and municipalities are defined by statute. This applies, for example, to building on development sites, land acquisition and the development of various communal facilities. Minor building projects, as a rule, are conducted solely on the basis of the legislation, but there are other development situations in which the statutory rules need to be elucidated and supplemented by voluntary agreements.

One type of agreement extensively used for land development is the development agreement (*genomförandeavtal*). This is an agreement at civil law between the municipality and a property owner or developer that defines common objectives and/or the rights and obligations of the parties in connection with a development. The agreement may specify what is to be built, when it is to be built, changes of landownership, who is to defray various development costs, and who is to be responsible for the construction of municipal engineering installations and for other measures. One form of development agreement is characterized by the property owner or developer owning the land that is to be built on when the agreement is entered into (*exploateringsavtal*). For the municipality, this type of agreement is primarily a means of controlling the implementation of a development. It is a document providing the municipality with a guarantee that development will conform to municipal intentions. The agreement can make clear the measures to be taken and paid for by the municipality and developer respectively in the course of the development. Through the agreement the municipality can also specify the structure of the building development and the chronological sequence of building operations. The land development agreement can also provide a means of minimizing municipal expense and involvement in connection with the development. Then again, a contractual agreement on changes of landownership and charges for municipal streets, water and sewerage systems, and so on can be more flexible than a formal, statutory procedure. For the property owner, the agreement may be a precondition of the municipality drawing up a plan for the land in the first place. The agreement, in other words, is something the developer has to accept so as to be able to build on the land. But the land development agreement need not be regarded solely as a necessary evil for the developer. Like the municipality, the developer needs to clarify in advance the thrust of development and the allocation of responsibilities and costs between the parties. It is also in the developer's interests to save time by securing agreements that will obviate unwieldy statutory procedures.

A land development agreement is always signed *before* the detailed plan is adopted because the municipal negotiating position is based above all on the possibility of refusing to adopt the plan if the developer is not prepared to conclude a land development agreement. It may seem odd that the municipality should be empowered to exploit its monopoly and official status to compel the developer to sign a land development agreement, but the possibility of a developer feeling constrained to sign is no impediment to such an agreement. The critical question concerns the terms of agreement that the developer can be forced to accept. In a land development agreement, the municipality must not exact measures or undertakings from the developer that exceed those the municipality could achieve by implementing the legislation regarding, for example, transfers of land and charges for municipal facilities. The "voluntary" signature of the agreement by the developer makes no difference in this respect. Of course, it can be difficult to draw a hard and fast line between permissible and impermissible terms of agreement, and in practice there has developed a grey zone in which it is accepted that developers will incur expenses or other commitments slightly exceeding what could be forced upon them by law (Hornsved 1991). It is quite clear, however, that this grey zone may not be extended into regular municipal taxation of the developer's profit on the land development.

Another form of development agreement is used when the municipality owns the land to be built on (*markanvisningsavtal*). This might be called a land allocation agreement. In connection with this agreement, the land is conveyed to a developer, and so the agreement can be viewed as a contract of sale combined with conditions to be met by the parties in connection with the development of the land. The municipal negotiating position when drawing up a land allocation agreement of this kind is based primarily on its ownership of the land. This means that a land allocation agreement can be entered into at any time, regardless of when the detailed plan is adopted; that is, the agreement can be signed either before or after the detailed plan is adopted. As regards the motives for drawing up a land allocation agreement, reference can be made to the purposes of the land development agreement. Perhaps the main motive of the developers is made clearer than ever by a land allocation. If developers do not sign the agreement, they will not have access to the land and will thus be unable to build on it.

As regards the demands that the municipality can make on developers, a land allocation agreement is not subject to the same legal restrictions as the land development agreement. Here the municipality acts as a landowner and vendor of the land. This makes the price, in the form of money or other types of consideration, that the developer is prepared to pay for the land a matter of negotiation.

A prior agreement (*föravtal*) precedes a subsequent development agreement. The prior agreement can be entered into for land owned by the developer or the municipality (or to be acquired by one of them). Thus, the prior agreement can lead either to a land development agreement or to a land allocation agreement. When a prior agreement is concluded between the municipality and the developer, no detailed plan has yet been drawn up. This is an important difference compared with land development and land allocation agreements, which normally are always geared to a prepared detailed plan.

In the case of land development or land allocation agreements, it is above all the municipality that is keen to have things cut and dried. In the case of a prior agreement, on the other hand, it is the developer who is most eager. Miller (1993) gives three principal motives for the developer entering into a prior agreement:

- When no detailed plan exists for an area, developers want to know what and how much they will be allowed to build. Developers are also interested in obtaining a preliminary decision regarding the development costs that they are to bear. These questions can be settled in an agreement of principle or a general agreement.
- Major development projects can involve various substantial investigation and planning costs. If so, it is an important question which costs are to be borne by developer and municipality respectively. It is also important to make clear who is to bear the risk of frustrated planning, i.e. the risk of the project being cancelled. Financial responsibility for investigations, planning and design work can be settled in a planning agreement.
- If the municipality owns the land, a potential developer may of course be interested in having first refusal of the land ahead of other developers. While investigating the feasibility of a project, the developer wants guaran-

tees of being able to buy the land later on if the project proves feasible. This can be arranged through an option agreement safeguarding the developer against competition during the investigation period and perhaps also regulating the price of the land.

Prior agreements present a legal problem. Through an outline agreement, the developer wants an undertaking from the municipality concerning the amount of building permissible in a development area. The municipality, however, does not have any real legal rights to furnish such guarantees at the stage of things when the agreement is concluded. Building rights cannot be determined until consultations concerning the detailed plan have been held with others affected by it. Option agreements present another legal complication. An agreement of this kind means a promise by the municipality to developers that they will be able to buy the land later on if certain conditions are met. The problem is that promises of future acquisition of real property are not legally binding in Swedish law. This being so, a prior agreement can be looked on only as a gentleman's agreement between the municipality and the developer. The parties pledge themselves to keep their respective sides of the agreement, barring unforeseen contingencies beyond their control.

Who benefits from the appreciation in land values in land development?

There are three main measures in connection with land development that have the effect of raising property values:

- planning and the award of building permits
- public investment in the form of schools, public transport, streets, water and sewerage networks, and so on
- the property owner's own investments in buildings and installations.

In addition, of course, there are underlying factors that give rise to appreciation: technical and economic development generally, population growth and other factors augmenting demand for a certain type of property in a certain area.

The question of who should benefit from the appreciation of land values resulting from a change in land use has been keenly debated for a long time. The present situation can be summarized as follows:

- The property owner obtains the appreciation resulting from his own investments in buildings and installations.
- Appreciation resulting from planning and the award of permits, i.e. through the detailed plan and building permits, accrues in principle to the landowner, but the municipality can make charges to cover its planning and building permit procedure cost. The basis of planning and building permit charges must be stated in a tariff.
- As regards appreciation resulting from public investments, the situation is less clear. Municipal investments in streets and green spaces serving the development area, and in water/sewerage and district heating systems, can be financed by levying charges on the property owners. On the other hand it is not possible, in principle, for public investments such as main roads, public transport, schools, day nurseries to be financed by means of direct

charges collected from the property owners. It should be added, however, that in practice property owners have contributed financially towards facilities of the latter kind. During the boom years of the 1980s, at least, development agreements were signed in which the developer agreed to bear a portion of these costs.

The municipality, then, has limited opportunity of turning to its own benefit the appreciation in land values in connection with the development. To increase its share of that appreciation, the municipality must act as landowner by acquiring land more or less in advance of its development. In this way the municipality will be in a better position to benefit from the appreciation resulting from development, either by selling land to a developer in connection with the development or else by retaining and managing the land after it has been developed.

Prevention of measures that impede current planning If work is in progress on special area regulations, a detailed plan or a property subdivision plan, the municipality may defer a decision on building permit applications until the planning work has been concluded. Building permit decisions, however, may not be delayed for more than *two years*. The same provision applies if the municipality has requested an expropriation permit for a property. That is, the municipality can defer the decision of a building permit application until the government has decided the expropriation claim, but not for more than two years after the building permit was applied for. Similar rules apply to decision-making by the Property Subdivision Authority under the Real Property Subdivision Act, the Joint Installations Act and the Utility Easements Act. Permits under these enactments may not be awarded if the property subdivision, communal facility or utility respectively would prevent or impede future planning.

Information systems

Sweden has a computerized title and property register (*inskrivnings- och fastighetsregister*), i.e. a cadaster. Its main purpose is to give publicity and legal security to ownership of real property, to mortgages and encumbrances and to survey procedures. The computer-register is in the process of being built up and covers almost the whole country. Properties not yet included can be looked up in manual registers.

The computerized register gives the property designation (*fastighetsbeteckning*) and the registered owner. The purchase price is recorded. If a new owner has applied for registration of title deeds (*lagfart*) but, for some reason, this has not been granted (dormant registration), this too is made clear. Mortgages on the property are entered at their nominal amounts, often with a note of the mortgagee. The register contains particulars of mortgaged easements and beneficial interests on the property. Where applicable, a note is made of distraint, bankruptcy and executive sale, continuing expropriation or pre-emption proceedings, disputed ownership, injunctions under the Health Protection Act, improvement orders for rental properties, compulsory management of rental properties, and so on.

The register also includes the area of the property and particulars of any official survey work formalities undertaken. Current detailed development and property subdivision plans for the property are shown, as well as other public provisions and constraints on land use. The assessed value of the property is shown, broken down into land and building. The assessed value is supplemented by an entry concerning the use of the property, e.g. a rental housing unit with dwellings or an industrial unit with an advance factory. The register will also show if the property unit is conveyed by site leasehold (*tomträtt*), and if so it gives the purpose of the conveyance, the ground rent and the interval for reassessing the ground rent.

The register contains references to documents held by local title deed registration authorities (*inskrivningsmyndighet*) and property registration authorities (*fastighetsregistermyndighet*). The title registration authorities mainly keep conveyancing documents and agreements on beneficial rights and easements. The archives of the property registration authorities contain original maps and other information compiled in connection with the formation, alteration, delimitation, and so on, of property units. In addition these archives contain legally binding plans and planning provisions, as well as documents on certain publicly decided constraints on the use to be made of an area. Each authority keeps separate register maps (*registerkarta*) on scales of 1:2000 to 1:10000 showing the geographical location of the property units.

The authority in charge of the computerized title and property register is the Central Board of Real Estate Data (CFD - *Centralnämnden för fastighetsdata*) in Gävle. CFD is also entrusted with a variety of other computerized activities. Like the local tax authorities, CFD can supply detailed property information collected in connection with property tax assessments, such as square metre dwelling and non-housing space. A list of property owners, showing all property owners within a limited area, is also obtainable, as well as computer-stored information about detailed plans, and so on.

The principle governing information retrieval from CFD is that most things can be accessed from terminals at local property register, title register, property subdivision and taxation authorities. Municipalities, companies, private organizations, and so on can subscribe to leased lines.

Property information is also obtainable from the municipalities. This applies to current plans and also to plans in preparation. In addition, the municipalities can supply information on water and sewerage rates, street connections, district heating, building permit requirements, environmental stipulations, and so on.

More detailed particulars concerning fixtures, mortgages, beneficial owners and tenancy conditions have to be obtained from the property owner direct. There is no official register of tenancies.

Register of organizations Certain particulars about companies (e.g. property companies) and incorporated associations (e.g. tenant-owner associations) are also obtainable from public registers.

Swedish law recognizes three main types of company: limited company (*aktiebolag*), partnership (*handelsbolag*) and sole proprietorship (*enskilt bolag*).

Particulars of limited companies are contained in the computerized company register kept by the National Patent and Registration Office (*patent- och registreringsverket*) in Sundsvall. The Office also maintains a register of incorporated associations and sole proprietorships and a register of co-operative economic associations and non-profit trading associations and foundations, in so far as they are obliged to maintain accounting records. The register is compiled by counties. Similar registers are being kept manually by the county administrations until all data have been computer stored centrally. The above registers give the name and organization number, address, line of business, signatory/ies and, in the case of limited companies, particulars of directors, managing director, auditor and share capital. They also provide information on any bankruptcies, and so on.

The registration authority for chattel mortgages (*inskrivningsmyndigheten för företagsinteckningar*) is located in Malmö and keeps a computerized register for the whole of Sweden, showing traders who have mortgaged their businesses, the mortgage amount, the mortgagee if known, and any measures of mortgage relevance taken by a creditor or public authority.

All the data registers mentioned above can be accessed from terminals such as the CFD terminals already referred to. In addition, the CFD is compiling a register of directors, signatories, and so on of joint property management associations, information that is otherwise obtainable from the property registration authorities.

3.2 The financial environment

Financial enterprises and institutions

Sweden's credit market (i.e. money and securities market) includes several institutions of importance to the property market, namely banks, "intermediary institutions" and finance companies (SOU 1988, 1989). The General Pensions Funds (*AP-fonderna*) and private insurance companies can also be included here. Outside the regulated credit market, there are undertakings of other kinds, especially property-renting companies. Loans can also be advanced by private individuals. Finance companies and institutions also include investment companies, securities funds and securities trading companies. It is unclear to what extent these enterprises are involved in the supply of credit to property owners, but they play an important part in the capital market by virtue of their trading and investment in shares, bonds and loans.

The banks are divided into commercial high street banks (*affärsbank*) and savings banks (*sparbank*). The commercial banks have a tradition of supplying enterprise with short-term credit. The savings banks have specialized in households and the housing sector. Parallel to the deregulation of the credit market, however, the lines of demarkation between the different fields of banking have faded, as have the boundaries with other credit institutes. Banks are alone in being entitled to maintain deposit accounts for the general public. In principle they are not entitled to acquire shares or real estate as investment assets. On the other hand they engage in financial activities through subsidiaries and holding companies.

The intermediary institutions (*mellanhandsinstitut*, usually *kreditaktiebolag*) normally have a specifically delimited line of business. They form a link between capital investors and ultimate borrowers, and they finance their activities mainly by issuing bonds. Long-term finance is the traditional task of the intermediary institutions. These institutions can be distinguished as institutes for housing (*bostadsinstitut*), local government (*kommuninstitut*) and corporate finance (*företagsinstitut*). Most credit companies are owned by banks. A few are fully or partly State controlled.

The housing institutes finance their activities mainly by issuing housing bonds, most of which are acquired by the General Pensions Funds, insurance companies and banks. These intermediary institutions used to have the exclusive right of advancing State-subsidized first mortgage loans for newly built housing (see the next section). The housing institutes are sometimes referred to as property institutes (*fastighetsinstitut*), the reason being that they also advance quite considerable credits for non-housing properties.

The finance companies (*finansbolag*) are companies created to specialize in new forms of finance that the banks have not been suited to deal with, such as leasing, promissory note credits, factoring (the purchase of trade debts) and credit cards. These companies deal with business undertakings and households and often specialize in a particular kind of operation. Property mortgage loans play a marginal part in their operations. The companies finance their business mainly by short-term borrowing. Ownership is heterogeneous and several of the larger companies are owned by banks or by industrial and commercial enterprises.

The property-renting companies (*fastighetsrentingbolag*) deal with the sale–leaseback of properties. These companies do not come under any specific statutory provisions, but a government commission has recommended that they be given the status of finance companies. Their activities are financed with bank loans or with other loans contracted in the money market. Many finance and property-renting companies are owned by banks, insurance companies or industrial concerns.

The assets of the investment companies (*investmentbolag*) are mainly financial, in the form of shares and securities such as loans, bills of exchange and receipts from companies in the group. Investment companies are not usually numbered among the financial institutions, although through their shareholdings they play an active part in the financing and restructuring of enterprise. They are owned by many private individuals and legal bodies.

The securities or mutual funds (*värdepappersfonder*) have large, well distributed holdings of bonds and shares held on the participants' behalf.

The securities trading companies (*värdepappersbolag*) are entitled to trade in financial instruments on their own behalf or that of others, and to procure and administer such instruments. In this connection they also have an extensive lending business.

The General Pensions Fund, administered by several managing boards, is the largest institutional investor in the capital market, with most of its assets tied up in loans to the State, municipalities or county councils or in housing bonds. A small proportion is invested in shares. This fund is part of the public insurance system and is mainly financed through social security contributions.

The insurance companies (*försäkringsbolag*) are extensively involved in capital management. This may include ownership of shares and of real estate, and interest-bearing investments and loans. Insurance companies invest in real property, advance mortgage loans and have large holdings of bonds (housing bonds for the most part) and shares. In 1991 there were just over 500 insurance companies, 400 of them being small local general insurance companies (see Table 3.1).

Table 3.1 Number of finance companies, 1991, and of private insurance companies, 1989. *Sources:* SCB (1993f) and SCB (1991a).

Category	No.	Comment
Commercial banks	117	Of which seven foreign, all with minor or no activity
Savings banks	101	
Co-operative banks	1	Later converted to a commercial bank
Intermediary institutions	23	Of which 11 housing credit institutes
Finance companies	158	Only those with net assets > SEK50 million included
Investment companies	44	
Securities funds	363	Banks are not included
Securities trading companies	21	
Insurance companies		
major	114	
minor	391	

The financing of property development (construction loans)

During the actual construction phase, housing production is financed with short-term credits known as construction loans (*byggnadskreditiv*). These are advanced mainly by banks and intermediary institutions, but also by finance companies. They are usually drawn on as and when funds are needed for the further progress of building. The loans carry market interest rates and security for them takes the form of mortgages on the property or, in uncertain situations, a municipal guarantee. When the building is completed and security for the loan is high, the construction credits are called in for payment and replaced by final credits, which usually have to be paid off in 40 or 60 years. If a developer sells the property, for example to an owner-occupier or a tenant-owner co-operative, the new owner takes over the final credits.

Non-residential development can be similarly financed, but here there is a wider choice of financial arrangements, added to which, complete or partial self-financing is probably more common. Industrial building development is often financed as a part of the company's general financial arrangements.

Infrastructure development, such as the construction of streets and water and sewerage mains, is financed either by the developer through construction loans or out of taxation revenue. If the developer is to incur expenditure in connection with such development, then this shall be stated in a development agreement. If the municipality is responsible, it can subsequently pass on the expense to the property owners by levying charges.

The financing of property ownership (housing)

For new housing production financed with State loans, the practice until 1991 was for the housing mortgage institutes to provide first mortgages (*bottenlån*) on the security of up to 70% of the property mortgage value. Depending on the type of ownership, State loans (*statliga lån*) were advanced for a further 22–30% of the mortgage value. The remainder was financed out of equity or with third mortgages (*topplån*) from credit institutes, usually banks or insurance companies. The State also makes grants towards interest payments on the first mortgage and State loans. To prevent the credit system resulting in cost inflation, a ceiling was imposed on permissible production costs, added to which, the costs of the individual project were subject to verification. Nearly 100% of all housing construction was being financed in this way at the end of the 1980s.

The structural principles of the system of housing finance as it then existed are illustrated in Figure 3.6. That figure also shows the function of the housing bonds market and the main interests involved on the financial side. The reason for illustrating the earlier system in this way is that the figure shows the structure and sources of earlier housing loans that are still being paid off. Eventually, however, both State loans and the former interest allowances will be phased out.

The foundations of the system were laid in the mid-1970s. There were quite a few changes of detail with the passing years, but the essential structure remained unaltered. Certain changes occurred in 1992. State housing loans were abolished, for example, and all finance is now contracted in the open market.

Completely new rules were introduced in 1993 with the aim of simplifying the system and rapidly scaling down the subsidies (Boverket 1993). Although, in all probability, the system will undergo quite a few changes in the future, it is worth describing here and now because it represents a break with earlier subsidization principles. The brief account that now follows is based on conditions in 1993. Credit flows, however, will probably continue to conform to the arrows in Figure 3.6, except that the loans are no longer divided into different types and can now be advanced by all credit institutes.

The State will continue to provide interest allowances (*räntebidrag*) for new housing production, but these will be unaffected by the manner in which building is financed in the credit market. Grants will be based on flat-rate computations of construction costs and interest expenses. The qualifying dwelling space is up to 120 m^2 per dwelling unit, and any building larger than this will receive no grant for the additional floor space. The computation base (*bidragsunderlaget*) is SEK13000 per m^2 for up to 35 m^2 and thereafter SEK6000 per m^2 up to 120 m^2.

For this standardized base, the State also fixes a standard interest expense. For the first year the State then provides 57% of the standardized interest expenses for rental and tenant-owner housing, and 42.67% for owner-occupied homes. The percentage rate of the grant then declines year by year, so that after nine years it will be 25% for rental housing and tenant-owner co-operatives and nothing at all for detached and semi-detached houses. The differences regarding the initial interest allowance and the speed of phase-out stem from the desire on the part of the State to equalize housing costs for different bonds of tenure. Because different rules of taxation apply to the different forms, parity is achieved by varying the interest allowances.

Thus, the grant for rental and tenant-owner properties during one year will be:

computation base × subsidized interest × 57%.

Although finance is raised in the open market, the State can, on payment of a special charge, give the lending organization a credit guarantee for 30% of the loans. This guarantee applies to the lender, not to the borrower. If the guarantee is used, the borrower will become liable to reimburse the State. The State can also advance redistribution loans to spread part of the annual interest expenses over more time. Heavy initial interest outgoings can be avoided in this way.

Loans and grants for the conversion of multi-family dwellings are based on the same principles as for new developments. The computation base refers to reasonable conversion costs for essentially improving the basic residential function of the building. However, the base may never exceed 80% of the base for a

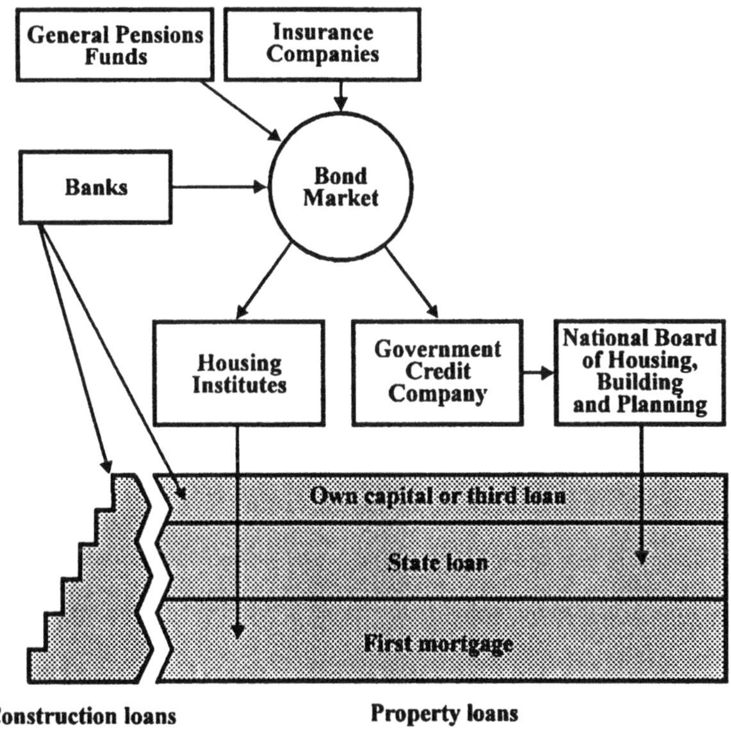

Figure 3.6 Credit flows under previous systems of housing finance (*Source:* SCB 1993a).

newly constructed building.

Non-residential and industrial premises are financed entirely without State subsidies.

Financing property acquisition with a mortgage loan

The purchaser of a property often needs to finance the acquisition with a mortgage loan. Loans of this kind are usually obtained from banks or other financial institutions. If the property is mortgaged, then in principle the seller has to pay off the loans. However, buyers can take the loans over if they so wish and if lenders allow them to do so. One of the parties may possibly demand renegotiation of the loans in this connection. The buyer may also choose to pay cash or to look for a new lender. The need for a new lender is especially noticeable when the buyer is unable to pay cash for a lightly mortgaged property.

The main principle is that loans may be fixed-term (*bundna lån*) or without fixed maturity (*obundna lån*). Fixed-term loans run for a certain period, usually two or five years, and at a fixed rate of interest. If property owners have to pay off a loan in advance, as a result of the property being sold, or wish to transfer the loans for some other reason, they have to negotiate with the lender on the conditions for paying them off, unless those conditions are already defined in the credit agreement. When the term of a loan expires, it is the conditions of the original agreement and the wishes of the parties that decide whether the loan is to be repaid or its conditions renegotiated, or whether the loan is to continue on the same conditions as before.

Loans without fixed maturity carry a variable market rate of interest and can be paid off at any time. The manner in which fixed-term loans and loans without fixed maturity are to be paid off is defined in the credit conditions.

Property leasing

Sale–leaseback transactions are exceptionally widespread in Sweden by comparison with the rest of Europe, and so these activities will be described in somewhat greater detail (further to this point, see SOU 1991). The popularity of these transactions has been attributable to a desire on the part of companies to release capital without damaging their balance sheets. Under this type of arrangement they retain the long-term use of the property, the control of it and the benefit of rising values.

Property leasing (*fastighetsleasing*) is not defined by law, but even so it developed rapidly in the 1980s. What happens is that the owner sells the property to the leasing company and then rents it back (hence the term, sale and leaseback), at the same time having an option of repurchase. The term property renting (*fastighetsrenting*) sometimes refers to transactions without sale and leaseback, with the parties drawing up tenancy and option agreements instead. The latter type of agreement is a marginal phenomenon compared with sale–leaseback transactions.

Tenancy agreements are drawn up in such a way that the rent consists of interest, depreciation, insurance premiums, property tax and any ground rents that may be payable. The interest portion can be computed after inflation plus real interest, usually about 5%, or entirely at market rates. The tenant arranges and pays for all the tasks normally incumbent on the property owner. The agreements normally run for between 10 and 24 years.

The sale and leaseback phenomenon was introduced on to the Swedish market by the SPP insurance company in 1978. The number of enterprises has grown since then. Their owners and participants mainly include insurance companies, banks and construction and property companies. The total property holdings of the property-renting companies at year's end 1990 were estimated at SEK35000 million book value. About 95% of their holdings are subject to repurchase options. Originally the renting companies addressed themselves mainly to large industrial companies and municipalities. Gradually they have come to act at all size levels and in all sectors. Their customers include small and medium companies as well as State-owned enterprises, municipal companies and county councils. The public sector volume, however, is estimated at no more than 10% of the total volume. The market consists of industrial, commercial and office properties, but also includes schools, sport halls, harbours and refuse disposal plants. Properties are scattered all over Sweden, although the main concentration of value is within the metropolitan city areas. Multi-family dwellings are seldom involved in this kind of transaction.

The properties are normally conveyed to limited partnerships, to ensure the feasibility of repurchase. Sale of the companies counts as a sale of personal property, which can be made subject to an option, whereas Swedish law does not countenance options on real estate. In 1991 the renting companies were estimated to own about 300 limited partnerships of this kind.

3.3 The tax and subsidy environment

Swedish taxation: a general outline of the system

Because of the extent and growth of taxation in Sweden, the rules of taxation have been perpetually changing. The largest reform of taxation hitherto was introduced in 1991. Its main concerns were to simplify the taxation system, by taxing different fiscal objects on a more uniform basis, and to reduce marginal taxation rates. During the 1980s marginal taxation for some persons exceeded 85% of net income (gross earnings less social security contributions), whereas the maximum rate of taxation under the new system is 50%.

Even since the new reform, further changes have been made to the taxation system, and new ones are planned. The system, in other words, is not very stable. Bearing this in mind, we will now tend to consider the rules applying in 1993. The reason for treating the taxation system as a whole is that it is intended to be generally applicable, and accordingly real property, with certain exceptions, is not specially regulated.

Viewing taxation in an historical perspective, one finds that direct taxation of companies and households has diminished in importance, whereas social security

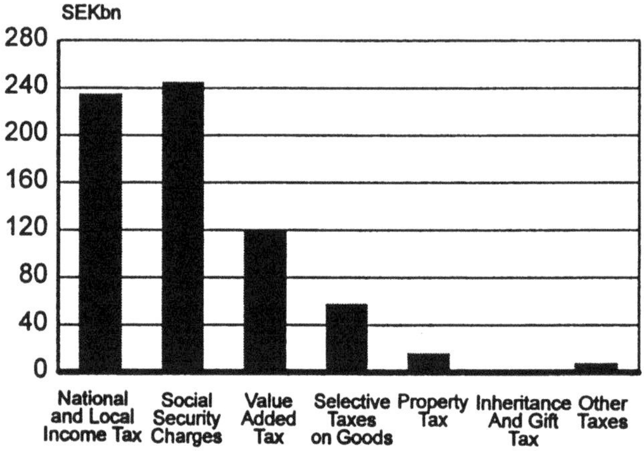

Figure 3.7 Taxes in the 1991/2 fiscal year (*Source:* Lodin 1993).

charges became increasingly important with the expansion of public transfer systems during the 1970s and 1980s.

The most important taxes can be classified as follows (wealth tax is not included because it was to be abolished after 1994): national and local income tax, social security contributions, value-added tax, selective taxes and charges on goods and services, stamp duty, property tax and inheritance and gift tax. The importance of these various taxes can be seen from Figure 3.7.

National and local income tax Both private persons and certain legal bodies are liable for taxation. Among the latter, limited companies and co-operative economic associations are special tax subjects in that first they are taxed on their operations, after which the owners are taxed on the dividend paid to them.

Income taxation for private persons refers to income from employment and economic activity (*förvärvsinkomst*), on which local and, for higher income brackets, national income tax are payable, and to income from capital (*kapitalinkomst*), which is separately computed and liable for national income tax only.

Income from employment includes wages, fees and other emoluments derived from regular or casual employment. Income from economic activity is taken to comprise the proceeds of independent commercial activity, e.g. consulting. Taxable earned income is arrived at by deducting expenses involved in the earning of income. There are also several general deductions that we need not go into here. Local and national income tax has to be paid on taxable income. The *local* taxation rate is on average 31%, comprising, approximately, 16% municipal tax, 14% county council tax and 1% parochial tax. There can be great differences, however, between individual municipalities. Actual taxation rates vary from 27% to 33%. *National* income tax is 20% of taxable income in excess of SEK186000.

Income from capital refers to revenue from property, such as interest income, share dividends and rental income, and also capital gains from the sale of property. Interest charges on loans can be deducted so as to arrive at the net yield. In the case of rental income, costs connected with the acquisition of that income are deductible. In the computation of capital gains, acquisition and improvement costs can be deducted from the proceeds of sale. Capital is taxed separately from earned income. If the computation of income results in a surplus, 30% national tax is payable on the income. If there is a deficit, a tax reduction is allowed of 30% of the deficit or 21% on deficits in excess of SEK100000. The basic idea is for all types of capital income to be taxed on the same terms, but in practice special rules have been introduced for real property and tenant-owner homes. These alternative rules can be used by the owner if they are more advantageous than the main rule of 30% tax on nominal capital gains.

The alternative rules for permanent homes, other private homes and tenant-owner homes have been introduced so as not to impede relocation. The capital gain here has been estimated, on a standardized basis, at 30% of the selling price. For a permanent home and tenant-owner home the tax is 30% of the profit, i.e. 9% of the sale price. For other private homes, e.g. weekend cottages, the taxation rate is 60%, i.e. 18% of the sale price. For commercial properties, on the other hand, the alternative rule means that the standardized capital gain is set at 90% of the sale price. Here again, though, the tax is 30% of the profit, which means that 27% of the sale price is creamed off as taxation.

Limited companies and co-operative economic associations (which are taken to include tenant-owner co-operatives, for example) pay only national income tax on their economic activity (30% of taxable income). The owners then pay tax on their dividend as income from capital (see above). In the case of partnerships (*handelsbolag*), on the other hand, profits, for taxation purposes, have to be included in the partners' taxable incomes.

Social security contributions An employer or company has to pay social security contributions ("employers' contributions") in relation to the wage bill. These contributions are also levied on people deriving income from economic activity. As from 1993, the employer's contribution equals about 30% of the wage bill. The original purpose of these contributions was to finance special benefits such as sickness allowance and pensions. Gradually, however, the contributions have been raised without a direct benefit linkage. Parts of the employer's contribution, in other words, can now be viewed as a direct wage tax.

Value-added tax Value-added tax is a general, national consumption tax levied on most of the goods and services produced in Sweden. It is paid by the producer of the goods or service, at a rate equalling 25% of the sale price. A reduced rate of 21% applies to foodstuffs. Certain goods and services, however, are exempt from VAT. They include, for example, medical and social care, pharmaceuticals and banking and financial services. More interesting for present purposes, however, is the VAT exemption of freehold and site leasehold property conveyances, leases, tenancies and tenant-owner titles.

Selective taxes and charges on goods and services In addition to VAT, certain commodities are liable for selective taxation. Special taxes are levied, for example, on petrol, energy, advertising, tobacco and alcohol. National and municipal authorities, as we have already seen, are entitled to charge for some of their services, including for example the preparation of detailed development plans and the award of building permits. Then again, property subdivision authorities finance their activities by charging for property registration.

Stamp duty The stamp (*stämpelskatt*) duty levied on the acquisition or mortgaging of a freehold or site leasehold property is another form of selective tax on property, payable to the State. Stamp duty on acquisitions of property applies to the purchase, exchange and compulsory acquisition (although not the inheritance or acquisition by gift) of property. For private and certain legal persons (e.g. a municipality, a non-profit housing utility or a tenant-owner housing co-operative), the tax is 1.5% of the value of the property. For legal bodies the taxation rate is 3.0% of value. Taxation is based on the purchase price or on the assessed value of the property (see below), whichever is the higher of the two. The stamp duty on cash mortgages equals 2.0% of the mortgaged amount.

Property tax Property tax is levied on detached and semi-detached houses, farmhouses and rental housing. Industrial and purely agricultural properties, accordingly, are exempt. So too, for the time being, are commercial properties and newly completed housing. As from 1993, no property tax is payable on commercial premises. If a rental property includes both housing and non-residential premises, then in principle the assessed value is subdivided with reference to the non-residential floor space. Newly built or completely renovated homes qualify for a reduced rate of property tax: no tax is payable for the first five years; for the next five years a 50% reduction applies. Leaving aside these exceptions, the main outlines of the taxation system are as follows.

Owners of private housing properties, detached, semi-detached or terraced houses pay property tax at 1.5% of the property's assessed value. Corresponding provisions apply to private homes held by site leasehold; that is, the site lessee pays a tax equalling 1.5% of the assessed value of both land and building.

A tenant-owner participates in a tenant-owner housing co-operative, whereas the co-operative owns the property where the tenant-owner lives. It is the co-operative that is taxed for the yield on the capital that the property comprises. As taxable income, the co-operative has to declare a standardized income equalling 3% of the assessed value of the property (which makes the true rental income tax free). In addition, the co-operative has to pay property tax at 1.5% of the assessed value if the property is a single-family dwelling or 2.5% if it is a multi-family dwelling.

Non-profit housing utilities come under the same rules as tenant-owner co-operatives. Income is 3% of assessed value and property tax equals 2.5% of the assessed value.

In the case of privately owned rental houses, actual rental income is taxed as income from economic activity. The provisions concerning property tax are the same as for other rental housing, i.e. the tax equals 2.5% of assessed value.

Assessed values are set by a process of general and special property assessment.

Nowadays, general property assessment comes in different years for different types of assessment unit. Rental housing properties, industrial, development and special-purpose units were assessed in 1988, single-family housing in 1990 and agricultural units in 1992. Each category will then be reassessed at six-yearly intervals. A special property assessment is made every year in cases where reassessment is called for, for example on completion of a building or extensive alterations. The basic principle is for the assessed value to equal 75% of the market value of the property two years before assessment. Thus, the 1992 assessment was based on prices in 1990. For developed properties, assessed value is divided into building value and land value.

Inheritance and gift taxation The same rules of inheritance apply to all property, real estate included. Inheritance tax is payable with reference to three classes. Class I comprises husband or wife or common-law spouse, children, grandchildren and children's spouses. Class II comprises other heirs or legatees not coming in class III. Class III includes churches, county councils, municipalities and non-profit associations of various kinds. A tax-free amount is first deducted from each inheritance portion. For classes I–III this amount is SEK280000, 70000 and 21000 respectively. Tax on the excess is computed according to differing scales, depending on the class concerned. For class I, for example, the scale of taxation is 10% in the SEK0–300000 band, 20% in the SEK300000–600000 band and 30% thereafter.

Tax, of course, can be computed only after the property has been valued. Where real estate is concerned, taxation is based on assessed value, not on market value.

Gifts are in principle tax free so long as their value does not exceed SEK10000. That limit applies to gifts to one and the same recipient during one and the same calendar year. Otherwise, gift taxation is subject to the same scales of taxation and the same property valuation principles as inheritance taxation.

Subsidies

Public subsidies to the land and property sector can be divided into two main groups:

- subsidies aimed at reducing the housing costs of tenants, tenant-owners and house owners
- subsidies paid to purchasers of developed land.

Housing subsidies The greater part of direct support for housing is channelled through State interest allowances for both new production and alterations (see Ch. 3.2). Table 3.2 shows total State expenditure on interest allowances between 1981 and 1991.

Apart from the general interest allowance, there are allowances of different

Table 3.2 State expenditure on interest allowances, 1981–91. Source: SCB (1992h: 160).

Financial year	SEK million
1980/1	5027
1981/2	7215
1982/3	8916
1983/4	9858
1984/5	11868
1985/6	10922
1986/7	14423
1987/8	13220
1988/9	13233
1989/90	19635
1990/1	23058

kinds for particular groups. In 1991 more than 300000 households received allowances towards housing costs (*bostadsbidrag*). Nearly half the recipients were single parents with children living at home. Housing allowances totalled approximately SEK5000 million. In 1990, more than 500000 pensioners received housing support (*kommunalt bostadstillägg*) and payments under this head exceeded SEK6000 million. Disabled persons can qualify for loans and grants towards the cost of necessary alterations for disabled access (*bostadsanpassningslån och -bidrag*). The grant can be up to SEK30000 per dwelling unit. Altogether more than 20000 units were adapted in 1991, when grant payments exceeded SEK600 million.

As was mentioned in the preceding section on taxation, interest expenses can be set off against income from capital. A deficit qualifies for a tax reduction equalling 30% of the deficit. This type of deficit is often incurred by owners of single-family dwellings, the reason being that interest outgoings on the house usually exceed the owner's income from capital. In cases of this kind the tax reduction could, in a sense, be regarded as an indirect fiscal subsidy, and tax reductions for owners of single-family dwellings in 1991 came to just over SEK13000 million.

Finally it should be added that support for the housing sector is a hotly debated topic at present, partly because of the parlous state of government finances. There is, consequently, every likelihood of retrenchment measures being introduced during the 1990s, but exactly how remains to be seen.

Subsidies for developed land In many cases the municipality is the original owner of land that is later to be built on by private persons or companies. At the same time the municipality is responsible for the construction and operation of streets and water and sewerage mains, for example. In many municipalities, developed land is sold at subsidized prices – that is, it is sold below the market price and/or the municipality does not levy the charges it is entitled to for the infrastructure (see Ch. 3.1). Subsidization of this kind occurs in municipal sales of land for both housing and commercial purposes, as a means of keeping housing costs down and encouraging business development in the municipality.

This type of municipal subsidization, however, can be expected to diminish in future. Municipalities are in financial straits and in great need of revenue. Then again, one discerns a change in ideological attitudes, to the effect that municipal authorities should not subsidize building development.

CHAPTER 4
The process

4.1 Price setting

Pricing questions can be illuminated with a schematic model of the development of values and prices during the development process. In this model, value development can be divided into several stages or phases (see Fig. 4.1; LMV 1993b).

In principle, property prices are governed by the market. Before notions of a change in land use are conceived, the value of current land use to properties is determined. This can refer to agricultural and forestry land or to land that has already been built on.

When changes in land use come to be expected, expectation values (*förväntningsvärden*) arise. Land not included in a detailed plan but affected by expectations of a change in land use is referred to in Sweden as undeveloped land (*råmark*, literally raw land). In some cases, expectation values can develop on the strength of more or less uncertain assessments of future development. In other cases they may be triggered by municipal declarations about development, combined with the preparation of a comprehensive plan. The steady gradient of the graph in Figure 4.1 is of course a simplification of reality. In practice the graph is affected by the "model" to which the anticipated development refers

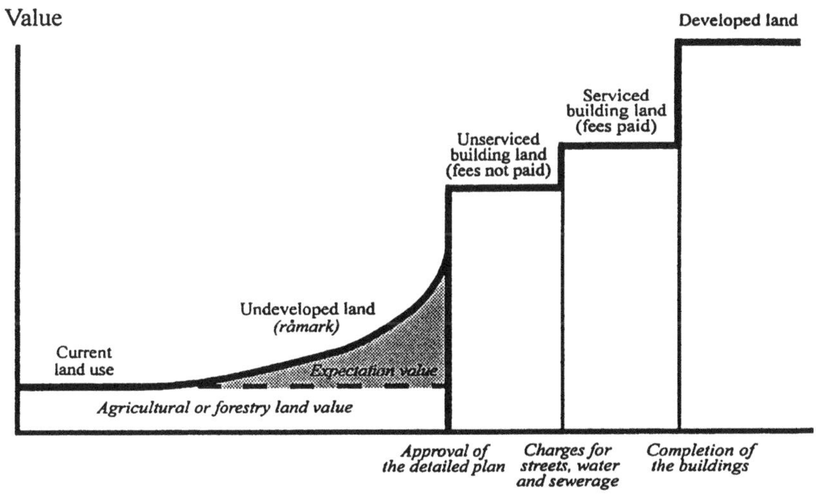

Figure 4.1 Movement of property values and prices during the development process (*Source:* LMV 1993a).

(see Ch. 4.3). In renewal and infill development of pre-existing residential and secondary housing areas with many property owners, value development normally proceeds in step with the detailed planning stages of consultation, exhibition and adoption. This type of case tallies relatively well with Figure 4.1 in which property values change little during the introductory phase of the process. Not until the detailed plan is approaching adoption do property prices display a more noticeable rise. If, on the other hand, we consider the case of a professional developer owning and intending to build on a large area, the situation is slightly different. Here the developer tries, as far as possible, to secure the development at an early stage, through agreements of principle or contracts with the municipality. In this case the planning process and the formal parts of the detailed plan come if anything as a confirmation of these agreements. Accordingly, the development of property values does not comply with Figure 4.1. Instead, the real rise in values comes earlier on in the process.

In principle, the development of expectation values is also governed by market forces. It should be observed, however, that many municipalities occupy a very dominant position in this market, in which case the municipality is a price leader (see Ch. 4.2). It should also be noted that the municipality is entitled to expropriate land for building development, under the Expropriation Act, at a price corresponding, in principle, to the value of the property according to current land use (expectation values accruing over the previous ten years do not necessarily qualify for compensation; see Ch. 3.1).

In connection with the detailed plan being adopted and becoming legally binding, complete certainty is established concerning the permissibility and nature of the development. The land is then upgraded from undeveloped land to undeveloped building land (*råtomtmark*). This elimination of the final traces of uncertainty concerning land use brings an instantaneous rise in value.

Apart from the value of undeveloped building land, the period following the adoption of the detailed plan includes another two phases or value levels. In order for plots in the plan to be built on, charges have to be paid for streets, water and sewerage installations, property subdivision, and so on (see Ch. 3.1). When this has been done, a value of "buildable" building land (*tomtmark*) accrues. This value level, of course, depends to a very great extent on the magnitude of the charges. As was mentioned in Chapter 3.1, the charges, in principle, may not exceed the cost of the installations or the measures taken. The third value level refers to the developed plot after building permission has been obtained and the building development on the plot completed. This final value level, of course, is "primary" in a development project and therefore governs the value movement occurring in earlier phases of the development process. This can give us cause to take a closer look at several factors influencing this value where different types of property are concerned.

The national system of housing credits

As we saw in Chapter 3.2, the national (State) system of housing credits has played a very important part in the financing and construction of housing. Until

1993, State loans were awarded on condition that dwellings met certain standard requirements and that the production cost or price did not exceed a specified level (which varied from one region of the country to another). The system of credits, in other words, incorporated a regulation of standards and prices intended for the resident's benefit. Consequently, the first-time conveyance of housing properties did not take place on a commercial basis. On the other hands no price controls applied, in principle, when the first owner re-sold the property. The system of national credits, however, has been radically transformed since 1993. A flat-rate State interest allowance, computed on certain grounds (see Ch. 3.2), is now paid, which among other things means that the former controls of standards and prices have been done away with. New housing output, in other words, has been market-adjusted.

Rules on rent fixing

The selling price of rental properties, housing and non-housing, is determined by the market. But at the same time, one very important factor determining the value of the property - anticipated rents - is to some extent regulated. The Property Act gives housing and non-residential tenants security of tenure (*besittningsskydd*), so that they cannot be instantly evicted. But, in order for this safeguard to have a real effect, rents also have to be regulated. In other words, the landlord must not be able to charge rents that are unfair.

The rules concerning housing rents lay down that the tenant cannot be forced to pay a rent that substantially exceeds the level of rents for similar properties in similar locations (*bruksvärdehyra* or utility value rent). For purposes of computing this utility value rent, the non-profit, municipal housing utilities are to be price leaders (see Ch. 4.2). The objective of this system is to have a restraining effect on rents, keeping them below the level of a notional market rent. Rent computation questions have been much debated during the 1990s. It has been queried, for example, whether the utility value system has created a reasonable relation between rents for properties in different locations and of different ages and standards. It is widely argued that differences between different properties have been insufficiently reflected by rents. Legislation for the market adjustment of rents was introduced in 1993/4 but was not enacted. Different rules apply, however, concerning the renting of commercial premises for shops, offices, industry, and so on. In these cases market rents, i.e. the rents obtainable in the open market, are permissible.

Tenant-ownership

Lastly, a few words about the pricing of tenant-owner (*bostadsrätt*) apartments. Under the earlier system of national housing credits, first-time conveyances used to be price controlled, whereas market prices could be charged on resale. There were cases in the 1970s and 1980s of attempts being made, by agreement with the tenant-owner, to regulate re-selling prices within a certain length of time, the aim being to prevent the first-time tenant-owner from deriving a rapid and large capital gain from the resale of the tenant-owner title, but these terms of contract were ruled invalid by the courts. Nevertheless, these changes in the system of national credits have led to a general market adjustment of the tenant-owner housing market.

4.2 The actors

The municipality

The municipality, as we have already seen, plays a pivotal role in changes of land use. It has many fields of responsibility with a bearing on the use of land resources, including education, child supervision, care of the elderly, social welfare, housing supply, energy planning, and environment and health protection. In addition, the municipality is directly involved in the development process in four other capacities:

- It is primarily the municipality that decides, through the planning and permit procedures described in Chapter 3.1, how land and water resources are to be used.
- The municipality has to cater for the need for certain forms of infrastructure (see Ch. 3.1). Briefly, the municipality is responsible for public places such as streets, green spaces and play areas. It is also incumbent on the municipality to provide and operate public water supply and sewerage installations and public district heating installations. Electricity supply for a development area is managed by special power distribution companies, many of them municipal companies or utilities.
- Many municipalities have substantial land-holdings. Basically, municipal landownership serves two purposes. First, the municipality itself needs land for public services, infrastructure and housing production (see below). Secondly, landownership gives the municipality a further means, in addition to planning and building permission, of controlling and influencing an impending development in various respects. The municipality often acts as a go-between by assembling strategic land more or less in advance of a development, land that it later transfers to a suitable property owner. This municipal role in the development process is illustrated by the case studies from Tyresö, Lidingö and Solna (see Ch. 6). Municipal landownership in connection with housing development has been very common. One contributory factor here has been the existence of certain rules attaching to the national system of housing credits. Under those rules, which applied until 1992, land must, in principle, have been in municipal hands in order for national credits to be awarded (*markvillkoret*, or the land proviso). However, these rules could be waived and, moreover, the provisions were not applicable to housing development on behalf of a private individual. On the whole, however, one generally finds over 80% of the land for greenfield development between 1972 and 1989 was transferred by the municipality to a final developer.

- The municipality or municipal companies can commission the production and management of building development, mainly for housing. The municipally owned non-profit housing utilities are a very important player in the rental housing market. By definition, the total rent receipts of non-profit utilities must correspond to their outgoings. As we saw in Chapter 4.1, the non-profit housing utilities are intended to set the level of rents charged by other property owners. Most of these non-profit organizations, roughly 320 of them, representing 900000 dwelling units and 1.5 million tenants, are affiliated to the umbrella organization SABO (the Swedish Association of Municipal Housing Companies).

Conditions vary to such an extent between Sweden's 286 municipalities that it is hard to give a more general picture of the municipal role. These differences have an effect, for example, on the scope and extent of building development. All municipalities, of course, share the aim of meeting local housing demand: all residents, in other words, must be entitled to good housing at reasonable cost. This aim, however, can be achieved in different ways in different municipalities. How are new homes to be divided between detached houses, terraced houses and apartment blocks? Are they to be conveyed freehold, on a rental basis or by tenant-ownership? How much of the building is to be done under the auspices of the municipal housing utilities? Then again, municipalities where the pressure of development is heavy can take different views of urban expansion and population growth. Some of them will try to curb an expansion, whereas others will welcome a growth in population. As regards economic aims, many municipalities aim to achieve full cost coverage in their development activities. Others, for distributive or other reasons, prefer to subsidize these activities out of municipal taxation revenue. Views on municipal landownership can also vary. By owning the land that is to be built on, the municipality can secure for itself the land value appreciation or profit that development entails. Partly for this reason, some municipalities try to acquire the land that is to be developed, primarily on a voluntary basis but in some cases by expropriation. Other municipalities are less keen on landownership, in which case most development occurs on privately owned land.

The State

The State, mainly through the county administrations, has the important task of supervising land use. In the course of the planning process, the county administration has to ensure that national interests are not encroached upon, ensure that building development is ecologically appropriate in terms of water and sewerage, noise, air pollution, traffic safety, fire hazards, and so on, and generally safeguard the interests of the State. In cases where land use planning involves more than one municipality, as happens for example in the metropolitan regions, it is the county administration that has to co-ordinate the interests of different municipalities.

As was shown in Chapter 3.1, certain questions of property law are decided by special national Property Subdivision Authorities. This applies, for example, to the formation and redistribution of property units, the establishment of communal facilities and the creation of utility easements in the form of the right to run pipes, and so on, over somebody else's land. It is also the Property Subdivision Authority that implements the Joint Land Development Act.

The State is also a major owner of property. Museums, castles and other historic buildings, as well as property owned by the Swedish State abroad are managed by the National Property Board. Akademiska Hus is a State-owned company responsible for higher education facilities. Vasakronan is a company that manages State-owned commercial premises, offices, and so on. Altogether Vasakronan owns 4.5 million m^2 of non-housing floor space in 4000 properties up and down the country, which makes it Sweden's largest property company. Today 90% of the tenants are national authorities and utilities, but Vasakronan's brief is to act like any other property company in the market, and a widening of the customer base is intended for the future. Mention should also be made of such State-owned enterprises as Telia, Posten and Swedish State Railways, all of which have considerable property holdings.

In addition, more or less against its will, the State has augmented its property holdings during the 1990s, as a result of Sweden's banking crisis. Total credit losses in the banking sector are estimated at some SEK175000 million, and a good deal of them emanate from property and construction credits (Ingves 1993). "Problem credits" in the State-owned banks Nordbanken and Gota Bank were transferred to two special companies, Securum and Retriva respectively. Securum administers credits in about 2500 properties and at mid-1993 owned about 500 properties, the plan being for the number to rise to about 1500 within a year. Most of these properties are commercial premises, valued at SEK10000–15000 million altogether. Through Retriva the State is expected to become the owner of about 500 properties having a combined value of some SEK5000–7000 million. The aim is for these companies to be wound up within a 15-year period, as and when the properties can be sold off at "reasonable" prices.

Property owners and developers

The property owner has two fundamental roles, namely that of changing the use made of the property by having buildings and facilities erected, and, subsequently, that of managing and maintaining the buildings and facilities constructed. Often these two roles are combined, i.e. the property owner carries out the development and then manages the buildings. It is also common, however, for a property owner to act as an intermediary by constructing buildings and then transferring them to new ownership.

In the construction of buildings and facilities that imply changes of land use, the property owner commissions a construction project. In this connection in Sweden, the term developer (*byggherre*) denotes a very heterogeneous category.

Developers might be private individuals who buy a plot on which to build a detached house for themselves. The case study in Chapter 6.1, concerning renewal of an older residential and secondary housing development, shows how some plots are thus developed by private individuals.

A developer can build homes that are subsequently transferred to a tenant-owner association and occupied by tenant-owners. Usually co-operative tenant-owner enterprises take charge of administration and of the co-ordination of building and management. Chapter 6.1 shows how HSB, the largest of these undertakings, commissions the development of a precinct. Other, similar organizations are Riksbyggen and Sveriges Bostadsrättsföreningars Centralorganization (SBC).

Chapter 6.2 shows how a private construction company builds homes for rental tenure. Chapter 6.2 also illustrates another common situation, namely that of a construction company erecting detached, semi-detached or terraced houses for sale. This situation is also described in Chapter 6.3.

Developers also include the property companies that erect housing or non-housing for letting. Mention has already been made of the municipal non-profit housing utilities, and Chapter 6.2 instances the construction of rental housing by a private enterprise. Sweden has about 45000 private property owners, and between them they control 65000 properties with 675000 dwelling units and the bulk of all non-housing premises. Roughly half these property owners, representing 45000 properties, belong to the interest organization the Swedish Federation for Rental Property Owners (SFF).

Chapter 6.4 shows an instance of a developer erecting premises for their own use, namely the construction of a new head office by Scandinavian Airline Systems.

Finally, the term developer (*byggherre*) also applies to a municipality constructing day nurseries, streets, water and sewerage mains, and so on. The same goes for other infrastructure mandators, such as the National Road Administration, power distribution and telecommunications enterprises, and joint-property management associations constructing communal facilities.

Contractors

Often the developer lacks the competence to engage directly in building operations. The companies carrying out the developer's "commission" in cases of this kind are called contractors (*entreprenörer*). A more detailed structural description of the construction industry was given in Chapter 1.4.

It is worth mentioning that many of the large construction companies are also large property owners. The largest construction company, Skanska, for example, is also one of Sweden's largest property owners. Most of its property stock has resulted from the retention and management of its own construction projects. In other words, the company has built on its own behalf by acting simultaneously as developer and contractor. The case study in Chapter 6.2 shows another of the large companies, JM Bygg & Fastighets AB, acting in this way.

Other property professionals

Building consultants In addition to contractors, a developer often needs the assistance of consultants of various kinds, for example planners, property economists and lawyers, architects, soil mechanics engineers, building designers, plumbing and electrical installation engineers and contract managers, responsible among other things for the co-ordination of planning, cost control, contract purchasing, building works and follow-up and inspection during the construction period. The possible role of the contract manager is described in Chapter 6.4.

Estate agents Single-family dwellings and tenant-owner apartments are almost invariably handled by estate agents (*fastighetsmäklare*), mostly on the seller's behalf. There are also other connections in which the services of estate agents have come to be more extensively used. The sale of a rental property, a single-family dwelling and a tenant-owner title through estate agents is described in Chapter 6.5.

Property economists and lawyers Consultants specializing in property economics work, for example, with investment appraisals of development projects, property management and valuation in connection with the sale and credit appraisal of properties. Similarly, legal advice is needed in connection with changes of land use and the conveyance and management of property.

Property managers Small companies or private property owners owning rental apartments or commercial premises seldom build up an organization of their own for the day-to-day management of their property. Instead they engage a company specializing in property management, all the way from financial control and accounting to direct maintenance operations (see also Ch. 1.4).

The building materials industry

The building materials industry plays a very important economic role in building, accounting as it does for something like one third of total construction costs. This industry is made up of many different branches: cement, lightweight concrete, building board, joinery products, plumbing and electrical products, and so on. Each of these branches in Sweden is characterized by few companies, some of which also command several sub-branches. As a result of vertical integration, construction companies like *Skanska* have also acquired interests in the building materials industry.

Financiers

The financial system already having been described in Chapter 3.2, it will be sufficient here to underline, once again, the radical transformation of building and property finance during the 1990s. There are several reasons for this. The very severe credit losses sustained by banks and finance companies, very much with reference to commercial properties, have of course made the credit providers extremely diffident about the award of new credits. At the same time, the national system of housing construction finance has been transformed. State housing subsidies are being phased out, the connection between State subsidies and housing standard requirements has been removed, the municipalities no longer vet credit and interest allowance papers, and there are no stipulations concerning municipality security for loans. In principle, then, credit assessments of housing construction projects have become the sole responsibility of the credit providers. Lastly, Sweden is passing through a profound and protracted recession, with great uncertainty about the future, and this is reducing demand for both housing and non-housing premises.

Consequently there are large points of uncertainty regarding the role to be shouldered by the financiers. How are they going to act in the long term and what will be the effects of their actions on land use and building? What influence will they demand in future on the location and design of buildings and facilities?

4.3 The process

The description of processes involving a host of activities, governed by several different sets of rules and official decision-making and involving many private and public actors, is of course a complicated business. Figure 4.2 gives an overview of the players, activities and regulatory systems in the development process. Although greatly simplified, it does convey some idea of the complexity of the process.

Many of these activities, actors and regulatory systems have been dealt with in previous chapters. One question that has not been raised, though, is the actual construction process, in which a project is finally realized and in which the main actors or protagonists are the developer and various contractors. We will therefore begin by describing this process before going on to describe four essentially different models of the Swedish development process, partly so as to provide something of a background to the case studies presented later on.

The construction process: contracts between developer and contractors

The purpose of this section is to clarify the terms and concepts occurring when a developer commissions a construction project from one or more contractors. First comes a basic description of the normal implementation of a construction project. This is followed by descriptions of various forms of procurement, i.e. different methods that the developer can employ when commissioning a project.

How is the order for a construction project placed? The principal stages of a construction project can be described as follows (see Fig. 4.3).

To begin with, developers (byggherre) may have certain requirements and preferences concerning the buildings and facilities to be constructed. Next they choose the appropriate form of procurement for the contract (see below). To give contractors a basis on which to submit tenders, there has to be a description of buildings and facilities in the tendering documents. This description may be purely verbal, but as a rule the documents include drawings of different kinds: architect's drawings, structural drawings, survey drawings, insulation drawings

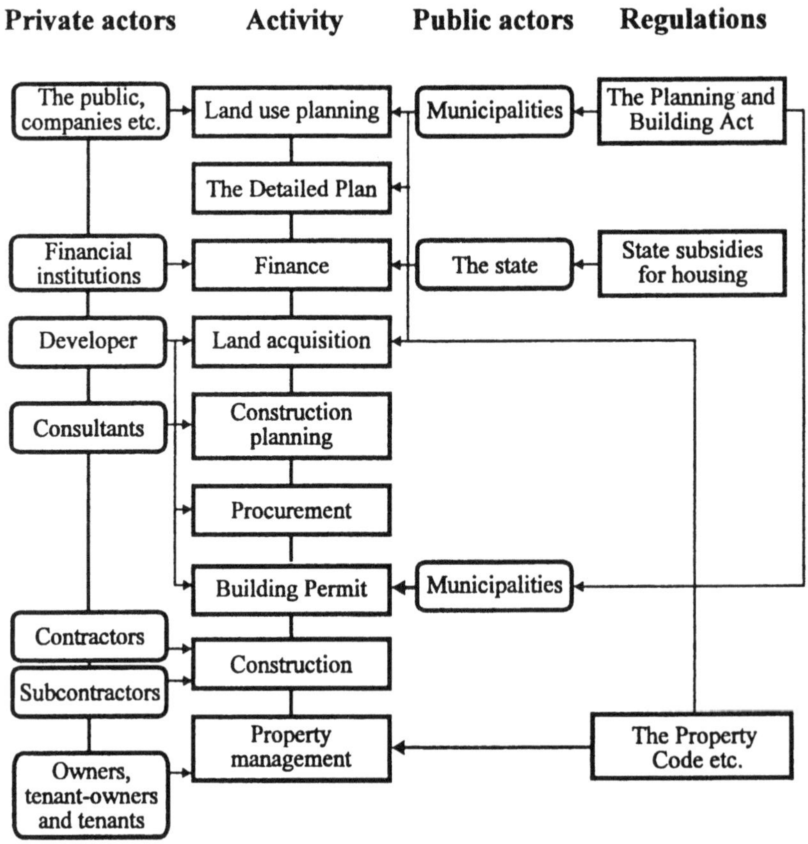

Figure 4.2 Actors, activities and regulatory systems in the development process (*Source:* SOU 1990c).

and wiring diagrams, and so on. Developers who lack the resources to prepare their own tendering documents must engage outside planners. After tenders have been invited and the tendering period has expired, the developer chooses a contractor and a contract is concluded between them. During the construction process or contracting period, the contractor in turn signs agreements with various suppliers and perhaps with subcontractors. When the project is completed, the developer checks that the contractor has discharged their obligations as defined in the contract. A final inspection takes place at the end of the contracting period and a guarantee inspection at the end of the defects liability period.

Forms of procurement The term "procurement" denotes the process from the invitation of tenders to the signing of a final agreement with a contractor (see

Liman 1983: 13). Procurement can take various forms, depending on the manner in which tenders are invited, the form of contract, and the form of price determination.

There are three principal ways in which developers can invite tenders from contractors. In a public or open invitation to tender, developers extend, usually by advertisement, an invitation to all companies interested in tendering. Limited tendering or tendering by invitation means developers turning to a smaller group of companies that, in their belief, possess the competence required for the work in question. Finally, developers may turn to a single contractor to discuss the terms of the construction project. This is termed "tendering combined with post-tendering negotiations". Open tenders, invited tenders and tenders with post-tendering negotiations accounted, respectively, for 13%, 20% and 67% of all dwelling units produced with State credits in 1991. Tendering with post-tendering negotiations showed a rising tendency in the 1980s (SCB 1993).

The form of contract governs the respective liabilities of the developer and

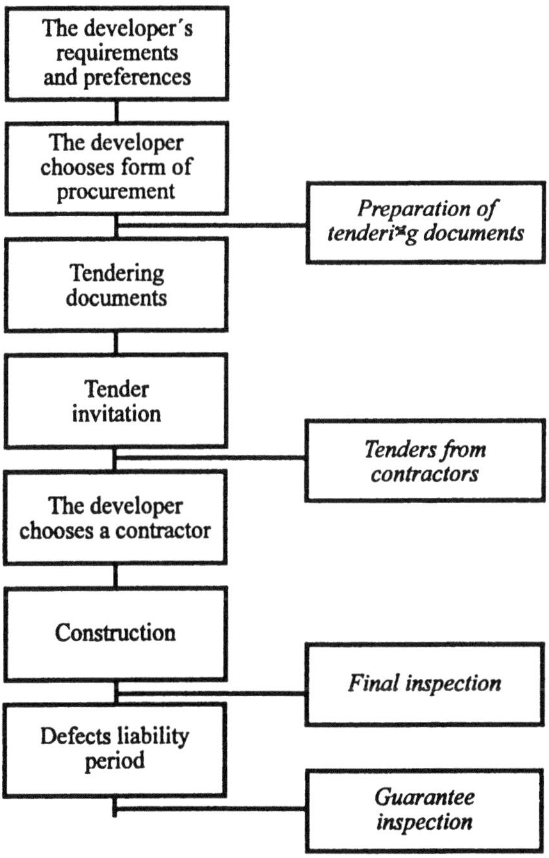

Figure 4.3 Model of the handling of construction projects (*Source:* Söderberg 1985).

contractor for different parts of a construction project. There are three main types of contract: divided contract, general contract and design-construct contract.

Divided contract Here developers, acting through their own personnel or outside consultants, take charge of planning. This form of contract is characterized by several contractors (the main contractor and joint contractors) being engaged by the developer. As a rule there are also subcontractors who have entered into agreements with the main contractor or a joint contractor. Divided contract means the developer taking charge of project co-ordination. In return, the developer can exert a great deal of influence on the construction process through the various contracting agreements. Another variant of the divided contract is for the main contractor to undertake the co-ordination of the project. This variant comes midway between divided contract and general contract.

General contract This represents an attempt to avoid the problems of co-ordination that a divided contract is liable to involve. In this case the developer signs an agreement with one contractor only, the general contractor, who in turn engages subcontractors. Here again, though, planning remains the developer's responsibility.

Design-construct contract This again is an agreement between the developer and one single contractor, but with the contractor taking charge of both planning and construction. In a "perfect" design-construct project, the developer's tendering documentation has a low level of detail, expressing functional requirements rather than cut-and-dried solutions. Sometimes, if the tendering documentation is purely verbal, reference is made to "programme purchasing". In certain cases, however, the developer has prepared "half-finished" plans and the contractor's influence is confined to matters of detail (Liman 1983: 19). Design-construct was the commonest form of contract in 1991, accounting for example for almost 70% of housing production that year (SCB 1993).

Forms of price determination There are two main forms of price determination in contracts: fixed price and cost-plus.

Fixed price means fixing in advance a price for all contract works. This may or may not be subject to index adjustment. If it is, the contractor will be compensated for general price rises during the contract period, and this is the procedure for the overwhelming majority of contracts. Another possibility is fixed unit pricing, in which case a certain price is agreed per volumetric quantity of the work done by the contractor. The reason for using this form of pricing is that sometimes, for example with earth-moving operations for a road construction project, it can be hard to know in advance how much work is involved. In a case of this kind, the price of the contract will not be known until the work has been done and the quantities determined.

Under a cost-plus agreement, contractors receive reimbursement for their verified expenditure on materials, instruments, work supervision, labour, subcontractors and certain kinds of insurance. They also receive a contractor's fee, to cover their administrative overheads and profit. The contractor's fee can take the form of a percentage of other expenditures, or it may be a lump sum. One possibility in a cost-plus contract is to define a cost ceiling for the contract, in which case any expenses over and above this will have to be borne by the contractor. A cost ceiling can be combined with incentive agreements, apportioning any shortfall below the ceiling between developer and contractor.

Legislation and trade conditions There are several laws and trade conditions governing contract relations. Legislation includes the Contract Act, the Sale of Goods Act, and legislation on trade restraints, unfair competition and marketing. In the trade context, general terms have been drawn up for various contractual relations. Provisions of this kind exist for building, heavy engineering and installation contracts, for the purchase of consulting services, for subcontracts, for design-construct contracts, for deliveries of building materials and for the construction of single-family dwellings. Tendering documentation is highly standardized. Descriptions accompanying drawings usually make reference to the various AMA books (AMA being the acronym for *Allmän material och arbetsbeskrivning*). AMA contains alphabetically and digitally coded descriptive texts to which developers can make reference instead of compiling a description of their own. AMA descriptions exist for buildings, site improvements and electrical and plumbing installations.

Four models of the development process

Kalbro (1992) distinguishes between different development situations on the basis of two factors with a vital bearing on the implementation of a development project.

- *Ownership conditions* within the development area – is the land privately or municipally owned when development is initiated?
- *The role of the developer* in the process – does the developer play an active part in work on the detailed plan together with the municipality, or is planning work entirely a municipal responsibility?

In terms of these factors, one can distinguish between four models of development project implementation (see Fig. 4.4).

	The developer does not participate actively in plan preparation	The developer and the municipality prepare the the Detailed Plan jointly
The developer owns the land	*Model 1*	*Model 2*
The municipality owns the land	*Model 3*	*Model 4*

Figure 4.4 Four typical models of development procedure (*Source:* Kalbro 1992).

Now it should be made clear that the development of an area need not necessarily conform entirely to any one of these four typical instances or models. A development area may include sub-areas that differ in character, in which case the development process may differ from one sub-area to another.

Model 1 In this case the land is owned by one or more developers who do not play an active part in drawing up the detailed plan other than by being consulted about it. This situation is typical of development in existing built environments such as renewal and infill development of older residential areas with property owners building their own single-family dwellings. A characteristic of these areas is the existence of several property owners, most of whom are not professionally involved in development and construction activities. Planning work and the construction of streets, green spaces, water and sewerage mains, and so on are therefore carried out by the municipality, the property owners' responsibility being confined to the construction of house and facilities on their own plots. The main outlines of the process in this type of development can be described as shown in Figure 4.5.

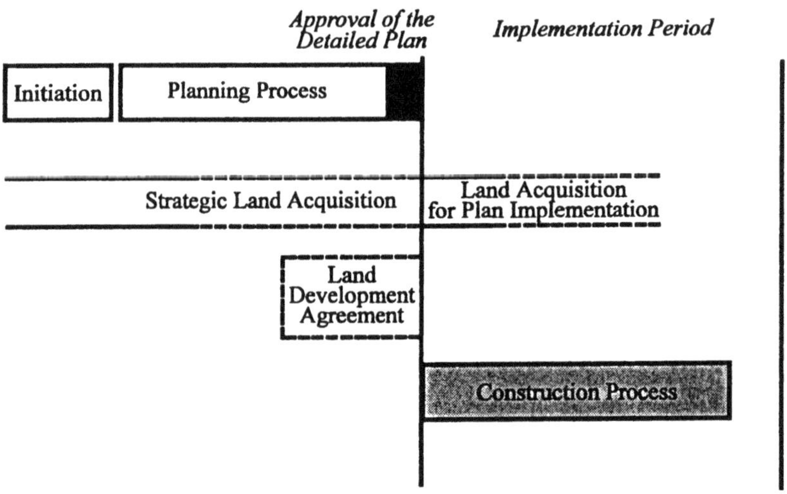

Figure 4.5 The development process: model 1.

Since the municipality decides where, when and how building is to take place, the formal initiative in land development is taken by the municipality through some form of decision in principle to go ahead with the project. The real initiative, of course, may also come from the municipality, for instance because of a need for housing, non-residential development or the enlargement and improvement of infrastructure provision. The initiative in changing the land use may also be taken by owners of a prospective development area. The owner or owners themselves may need new development on the property or intend selling the property or parts of it when the development is completed.

The planning process has already been described (see Ch. 3.1 and Fig. 3.5 above). The Planning and Building Act lays down that the detailed plan should be based on a programme containing a description of the present situation, preconditions for building development, the purpose of the planning, the timing of the planning process, and so on. The planning programme is not obligatory, because in many cases the same purpose can be served by the comprehensive plan, especially if the latter has been "deepened" or made detailed for the area to be planned in detail. As part of the planning preparations, various technical investigations are carried out with regard to geotechnical conditions, water and sewerage, roads and traffic, energy supply, and so on. Pre-planning of streets, services and other installations takes place. Planning design also has to be coordinated with building, economics, finance, land acquisition and possible implementation agreements and organizational questions affecting implementation of the plan. One purpose of the implementation report, which has to be drawn up in connection with the detailed plan, is in fact to force people to have regard to the plan in relation to other crucial aspects of the development process. In cases where a subdivision plan needs to be drawn up, this can be done concurrently with detailed planning. It should be noted that the subdivision plan is mainly of use when developing areas with several property owners, because it regulates property subdivision, communal facilities and utility easements, i.e. the legal relationship between property units. In areas with one developer only, questions concerning property subdivision, communal facilities, and so on can usually be clarified and settled in a development agreement between the municipality and the developer.

Figure 4.6 is a simplified description of the construction process already dealt with. It shows the order of activities for a divided contract and general contract when the developer, with or without consulting assistance, carries out the planning and then puts the project out to tender. The tendering documentation for a design-construct contract, by contrast, is less detailed, often taking the form of a verbal programme. Once the developer has appointed a design-construct contractor, the latter will be responsible for both planning and construction. In the ideal case, the contractor is selected on the basis of competition, but developers can also approach a single contractor or carry out the project under their own auspices.

In principle, the design of facilities and buildings begins when the detailed

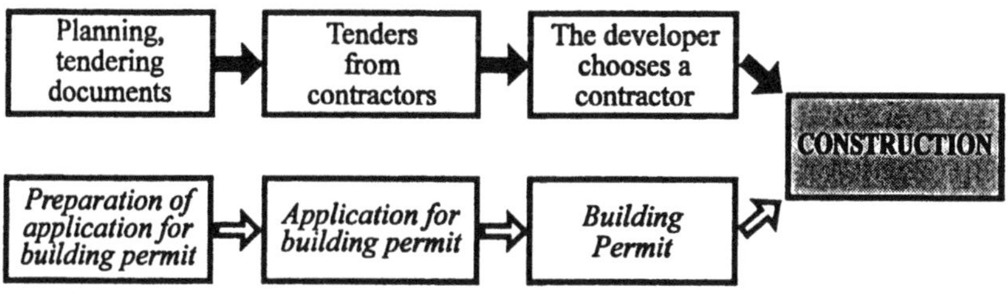

Figure 4.6 The construction process.

plan has acquired the force of law, i.e. when the content of the plan is known for certain. If the detailed plan is non-contentious, one can already be fairly sure during the final phase of the planning process what the eventual plan is going to look like. In cases of this kind, design can begin earlier but, of course, the actual construction of buildings and installations cannot begin until the plan has acquired the force of law and building permission been granted.

Primary responsibility for the construction process can vary somewhat with this development model. One option is for the developer to erect buildings and facilities on a development site, while the municipality attends to the construction of public spaces and communal water and sewerage mains. This is a natural allocation of responsibilities if the developer is not a professional builder, as for example in the case of a property owner in a renewal district. If, on the other hand, the possibility exists of the municipality and developer entering into a land development agreement (*exploateringsavtal*), then the agreement can lay down that the developer is to be responsible for constructing the municipal facilities and will then transfer them to municipal ownership.

If the developer owns the land, a land development agreement can be concluded between the municipality and the developer *before* the detailed plan is adopted. This, however, is subject to the development being on a certain scale. Developers do not have to own all the land in the development area, but in practice the law does require them to be in a position to sell "several plots". This means that development agreements can occur in model 1. But it is also conceivable that the prerequisites of an agreement may be lacking, for example, with the intensification of a single-family housing area if the original property units are to be subdivided into two or three new plots at most.

In Chapter 3.1 a distinction was made between strategic land acquisitions and land acquisitions for planning implementation, i.e. acquisitions made, respectively, before and after the detailed plan. Strategic acquisitions, made with a view to developing the land when the plan has been adopted, can be made before work on the plan even begins. Often, though, the commencement of detailed planning conveys a signal to the property market so that potential developers, whether construction companies or private companies, acquire land, while existing property owners with no interest in building dispose of their properties. Later, when the detailed plan has been adopted, land acquisitions are made with a view to adjusting property subdivision and ownership conditions to it.

Model 2 Here, as in the preceding model, the developer owns the land. One vital difference, though, is that municipality and developer work out the detailed plan together. Thus, the developer becomes involved at an early stage of the development process. This form of development can be described as in Figure 4.7.

There are two main reasons for the developer to play an active part in planning:

- One purpose is to improve the quality of the plan with regard to building development, economics, and so on. Co-operation between municipality

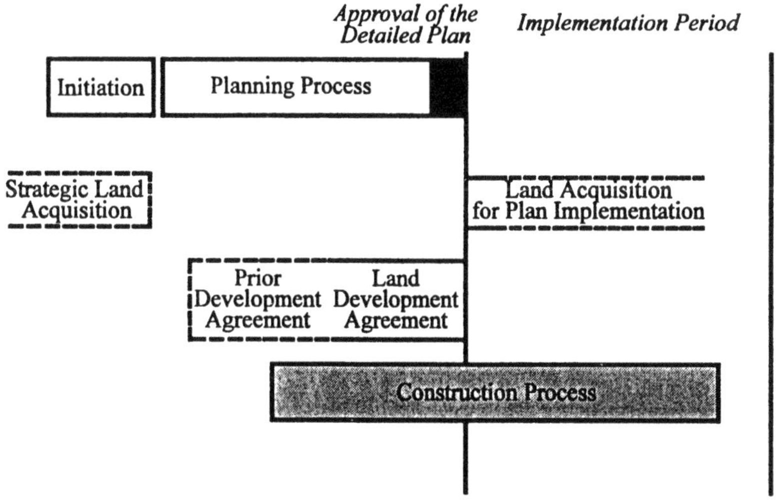

Figure 4.7 The development process: model 2.

and developer means that the planning and construction processes can be properly co-ordinated. The benefits of early developer participation are greater in the case of more extensive and complicated development projects where the municipality alone has difficulty in defining the requirements for the end product.

- If parts of the construction process proceed parallel to compilation of the detailed plan, this has the effect of speeding up the development process.

To be able to participate in the planning process, the developer must of course have the competence required. This competence is possessed by developers doing their own building, e.g. construction companies, or by developers who can engage consultants.

There are great similarities between models 1 and 2, and so the following remarks are mainly confined to essential differences between them. The practical business of drawing up the detailed plan can be variously organized. In cases where the municipality has a carefully thought out and politically accepted detailed planning programme, much of the practical work of planning can be entrusted to the developer. If so, the main task for the municipality will be to scrutinize the developer's planning proposals and indicate any changes that may be needed. In addition, the municipality must attend to the formal processing of the planning transaction including consultation, display and adoption. Another possibility is for the municipal authorities to assume a greater share of responsibility for the practical business of preparing the plan, whereas the developer takes principal charge of design.

One characteristic feature of this model is that much of the developer's design work runs parallel to the compilation of the detailed plan. If the plan is non-controversial, work on purchasing the construction project can also begin during the final phase of the planning process. The construction of facilities and buildings, however, may not begin until the plan has been adopted and acquired the force of law and the developer has obtained building permission.

When all the land in a development area is owned by one developer, almost invariably a land development agreement is drawn up between the developer and the municipality. This may be preceded by a prior development agreement. The agreement concluded before planning and design begin can define the tasks of the developer and municipality in connection with planning, technical investigations, apportionment of investigation and planning costs, timetabling, and so on. Another important point that may also be included in the agreement concerns cost liability for frustrated design or abortive work in the event of the project being abandoned.

In model 1, strategic land acquisitions could be made after detailed planning work has begun. Planning, in principle, is unaffected by landownership, since the developer does not play an active part in the planning phase. If, on the other hand, the developer is to participate in planning, a different situation applies. For obvious reasons, the developer must have acquired the land before planning begins. Land acquisition for planning implementation after the detailed plan can be regulated in the land development agreement. In that agreement, municipality and developer may agree that land for streets and green areas is to be conveyed to the municipality and that the developer is to reserve space for services, communal facilities, and so on, and they can agree on the payment to be made and on when and how property subdivision is to be effected.

The Joint Land Development Act Model 2 can also be said to include joint land development (see Ch. 3.1). Joint land development can be applied in areas with several property owners, i.e. areas where conventional development procedure normally requires model 1 to be used. Joint land development, however, makes it possible to switch from model 1 to model 2 (see Fig. 4.8).

The procedure for joint land development can be schematically described as follows. The process begins, formally, with the municipality adopting special area regulations (*områdesbestämmelser*). These regulations have to indicate the principal boundaries of the joint development area and the period within which the development decision has to be made. This decision is made by the Property Subdivision Authority on application being made by property owners or the municipality. Among other things, this decision has to show the property units that will be taking part in the joint enterprise, any stipulations concerning the development procedure, and the participatory shares (*andelstal*) regulating the financial apportionment between the property units. Through the development decision the property units form a joint land development unit (*exploaterings-samfällighet*), which is a legal entity. In this way the property owners are coordinated in an organization that can enter into financial commitments and conclude agreements. In other words, the joint unit can act as *one* landowner in the area, capable of engaging consultants and taking responsibility for design, and so on, as well as collaborating with the municipality in the preparation of the detailed plan. To provide for implementation of the development following the

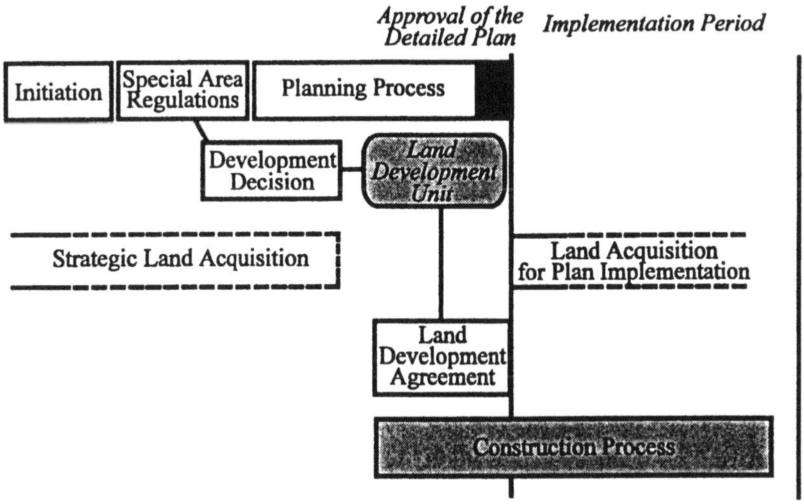

Figure 4.8 The development process: joint land development.

detailed plan, the municipality and the joint land development unit can sign development agreements, which is not normally possible with the different property owners acting individually. In the development process, the joint land development unit can take charge of design, procurement and construction of accessories such as streets, water and sewerage mains and communal facilities. On the other hand, the joint land development unit may not erect buildings on private development sites. That task devolves on the individual property owner. Strategic land acquisitions before the detailed plan are in principle possible until the plan is adopted. There is nothing to prevent property owners from selling their property, since participation in the joint land development unit is linked to the property, not to the owner. Acquisitions for implementation purposes after the detailed plan are made through land conveyance and/or reallotment. These conveyances can entail a relatively drastic restructuring of property units on development sites.

Model 3 In this instance the land is owned by the municipality, which also draws up the detailed plan. Once the plan is drawn up and adopted, the municipality appoints a developer through land allocation (*markanvisning*; see Fig. 4.9).

Because the land is municipally owned and the municipality is responsible for planning, the municipality will be sole agent during the introductory phase of development. That is, in this case the initiative for development will come from the municipality. Usually, however, that initiative is taken in light of the existence of potential developers who are willing to build in the development area. These may be companies that themselves are in need of housing or non-housing premises, or construction companies wishing to engage in speculative building.

When the detailed plan is drawn up, the municipality does not have a developer to consult and co-operate with. This means that the municipality must draw

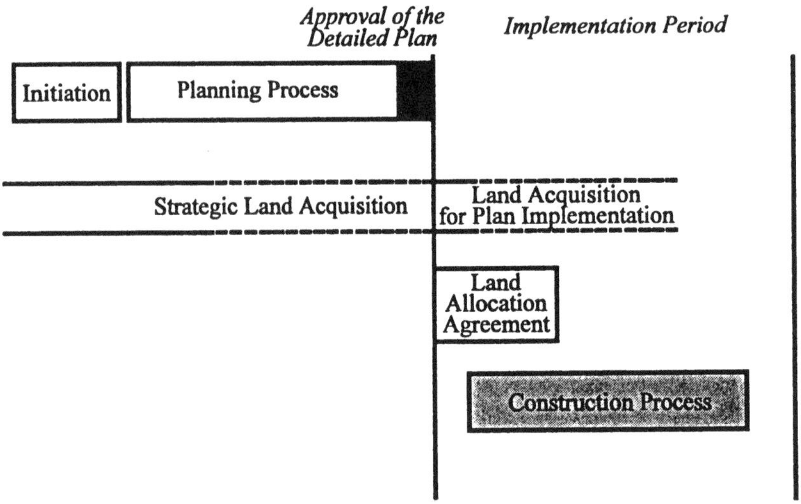

Figure 4.9 The development process: model 3.

up the detailed plan without any closer knowledge of the future developer's requirements and preferences. One way of overcoming this problem is to make the plan less detailed by restricting it to the minimum requirements laid down by the Planning and Building Act. The plan must at least show the use to which the land is to be put (housing, industry, transport infrastructure, and so on), the permissible right of development, the principal locations of services, parkland and traffic, the implementation period for the plan and, in the case of residential areas, single- or multi-family housing, type of building, maximum number of storeys, access, parking and main accessory spaces. Plans of this kind may, for example, be used in land allocation competitions for greenfield areas, competitions in which the more detailed design of the plan is an important competitive aspect (see below). It should be added that, normally, the original "flexible" detailed plan is altered and adapted to the winning entry.

The municipality can award its land to the developer or developers of its own choosing, whether non-profit, co-operative or private. Usually land is allotted to different developers in accordance with a quota between different developer categories, based on a certain order of priority or on other politically defined criteria. The municipality may also allocate the land after a land allocation competition, in which it invites developers to submit proposals for the development of the land. The developer making the best offer, for example in terms of quality and cost, is then awarded the land. Figure 4.10 shows how a land allocation competition can be adapted to the detailed plan.

The agreements reached by the municipality and developer in connection with an allocation of land are confirmed in a land allocation agreement (*markanvisningsavtal*). This makes provisions concerning acquisition of land, the price of the land and other financial questions, the design of buildings and facilities, responsibility for the completion of different measures, and so on.

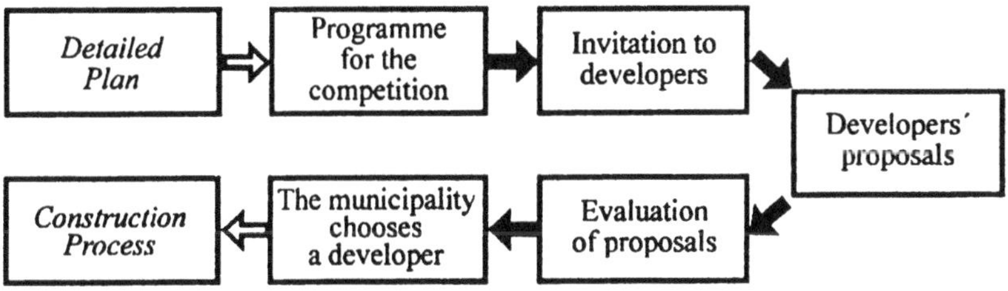

Figure 4.10 Land allocation competition after adoption of the detailed plan.

Model 4 In this model the land is owned by the municipality, which appoints a developer at an early stage of the process. After this the municipality collaborates with the developer in drawing up the detailed plan (see Fig. 4.11).

This model can on the whole be commented on by referring to previous models. The developer becomes involved at an early stage of planning, which means that initiation and compilation of the plan comply with model 2. Planning refers to municipally owned land; in other words, land can be acquired in the manner described in model 3. The land allocation aspect, however, must be elaborated somewhat. The developer is appointed before work on the detailed plan begins. The municipality may at that time choose whichever developer it finds suitable or else employ a competitive procedure. Figure 4.12 shows how this competition can be made part of the process.

When the developer has been appointed, an advance agreement (*föravtal*) is often concluded between the developer and the municipality. Among other things, this agreement may define responsibility for continuing planning and design work and the allocation of costs between municipality and developer.

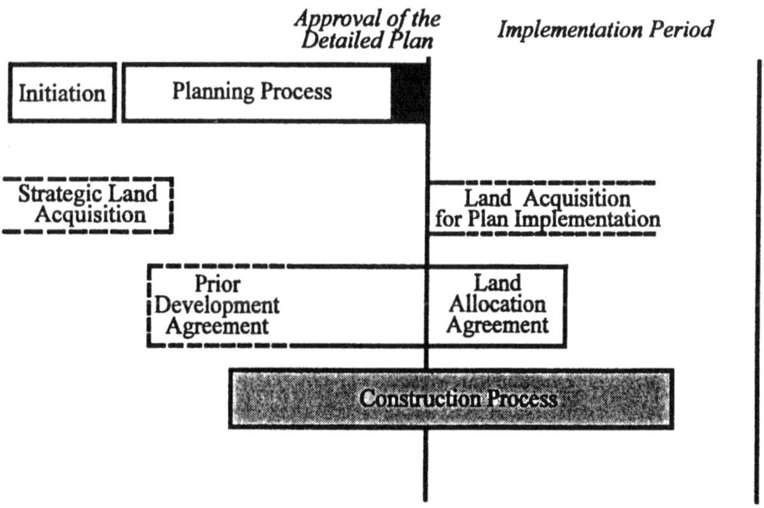

Figure 4.11 The development process: model 4.

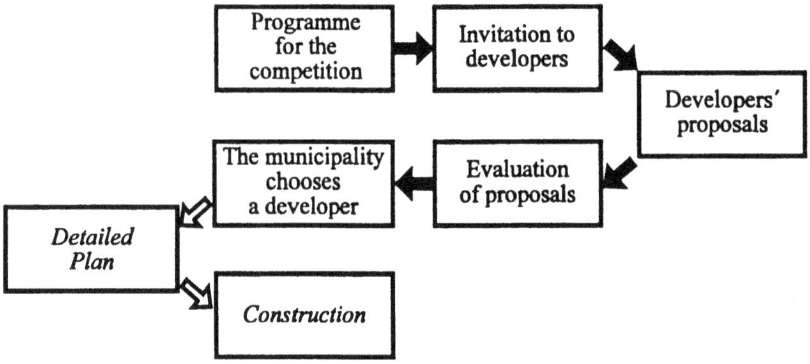

Figure 4.12 Land allocation competition prior to adoption of the detailed plan.

When the detailed plan has been completed, the final land allocation agreement is drawn up, to settle land transfers, financial questions, responsibility for different measures, and so on.

CHAPTER 5

The urban property market

The property market will be described with reference to the following submarkets: agriculture and forestry, building land, single-family dwellings, second homes, rental property (housing and non-housing) and industrial property. For each submarket, an account will be given of stock and owners. For housing and non-housing, age and quality will also be reported. Turnover, appreciation and depreciation, and price structure are considered. The last section deals with tax-exempt properties, although this is really a type of property without a market. Undeveloped land is dealt with in connection with land for agriculture and forestry. Site leasehold has been given a section of its own. In the background description of the Swedish property market (Ch. 1–2), the past 8–10 years were shown to have been a period of economic turbulence, a point to bear in mind when reading this chapter.

Purchase price statistics and purchase price coefficients The description of value developments in the property market is extensively based on official purchase price statistics. These are mostly presented in the form of "purchase price coefficients", a term that requires definition.

Particulars of all sales of real property are officially registered, as are purchase prices. All land, plus buildings, is assigned an assessed value, which is supposed to equal 75% of the market value at the time of assessment. The ratio between the actual purchase price (*köpeskilling=K*) when a property changes hands and its assessed value (*taxeringsvärde=T*) is called the purchase price coefficient (*köpeskillingskoefficient=K/T*).

Statistics Sweden (SCB) collects figures concerning, for example, these purchase prices. The purchase price coefficients are calculated for all property sales, and a number and variety of averages are then computed on an annual basis. Normal or "straight" purchases (those deemed to represent a market value) are distinguished from in-family purchases and other purchases that cannot be deemed to represent a market value. The codified results are published for the country as a whole, but also for counties and larger regions. As the statistics are published continuously, price movement time series are obtainable. Coefficients are published for agriculture and forestry, single-family dwellings (permanent and secondary), rental properties (housing and non-housing) and industrial properties. Statistics are also published for building land. The coefficient time series presented in the following sections have been adjusted to the value of 100 for 1981. Deflation to fixed prices has been carried out with the aid of the retail price index.

5.1 Land for agriculture and forestry

As we saw in Chapter 1.4, the land utilized for building or other urban purposes totals 320000ha (3200km^2). Previously undeveloped land totalling 17000ha (1ha = 10000m^2) was utilized between 1980 and 1990 (SCB 1992g). This represented a 5% urban expansion in ten years, equalling about 600 agricultural and forestry properties having an average acreage of 30ha, a moderate expansion compared with developments in the 1960s and 1970s, when the urban acreage increased by 20% per decade (SCB 1993h).

Thus, the average annual rate of expansion in the 1980s was 1700ha, corresponding to the utilization of 60 average property units for the country as a whole. This expansion, however, did not necessarily occur on newly acquired land; instead, the land may have been purchased by municipalities or other developers in earlier periods of expansion when land consumption was heavier. Municipal acquisitions, for example, were very extensive during the culmination of building development in the 1960s and 1970s.

Because urban expansion mostly takes place on agricultural and forestry land, the price picture for near-urban land of this kind is indirectly related to the urban property market. The lowest price of undeveloped land is governed by the prevailing agricultural and forestry value. In addition, the low demand for urban land is likely to depress the prices of undeveloped land in the direction of current agricultural and forestry land prices.

Another kind of pressure on prices exists in the provisions of the Expropriation Act. When land is expropriated, the appreciation occurring because of the expropriation enterprise must in principle accrue to the succeeding owner, not the seller. The purpose here is that no payment should be made for anticipated values relating to a change in land use. As a result of these provisions, undeveloped land values are influenced by the value of current land use rather than that of the alternative use for urban purposes, because as a a last resort the municipality can expropriate when a change in the use of the land is intended. The municipality can avail itself of this ability, even if the intention is for the plan subsequently to be implemented under private auspices. In that case the municipality becomes an intermediary in a chain of selling. Expropriation, however, is uncommon, because the players in the market know the rules of the game. Even so, the linkage of the price mechanism with the provisions of the Expropriation Act makes knowledge of agricultural and forestry values useful when purchasing undeveloped land, even if, in practice, both purchase and expropriation involve a certain amount of compensation for anticipated values.

In 1991 there were 275000 agricultural and forestry undertakings, of which 25000 comprised arable land only, 175000 forest only and 70000–75000 were a combination of the two. The combined undertakings owned about 4 million hectares of forest, an area that has diminished over the years, suggesting that they have tended more and more to specialize in either farming or forestry. The agricultural undertakings were owned and worked almost exclusively by private individuals. Roughly 50% worked their own land only, 35% worked both freehold and leasehold land, whereas the remainder worked leased land only (SCB 1993g,k). The term "undertaking" here refers to management units, for which reason the number of undertakings in Sweden does not entirely agree with the number of tax assessment units (see Ch. 1.4) or with the number of property units.

The average acreage of exclusively agricultural undertakings has been estimated at about 30 ha. So too has the average acreage of privately owned forestry (LMV1993d). This figure does not include State and corporate forest holdings. The average combined undertaking has a slightly larger forest acreage, upwards of 50 ha, plus arable land. Properties of this kind are much more common in the northern parts of Sweden.

Figure 5.1 shows agricultural and forestry acreages in Sweden in the 1980s, as well as non-productive land, by ownership categories. The extensive forest holdings owned by the State have diminished lately by privatization of the ownership and then through selling off of shares in the privatized company. The State however still owns half of the shares.

Every year in the 1980s, about 7% of the total number of agricultural and forest properties changed hands, and of these about 30% came on to the open mar-

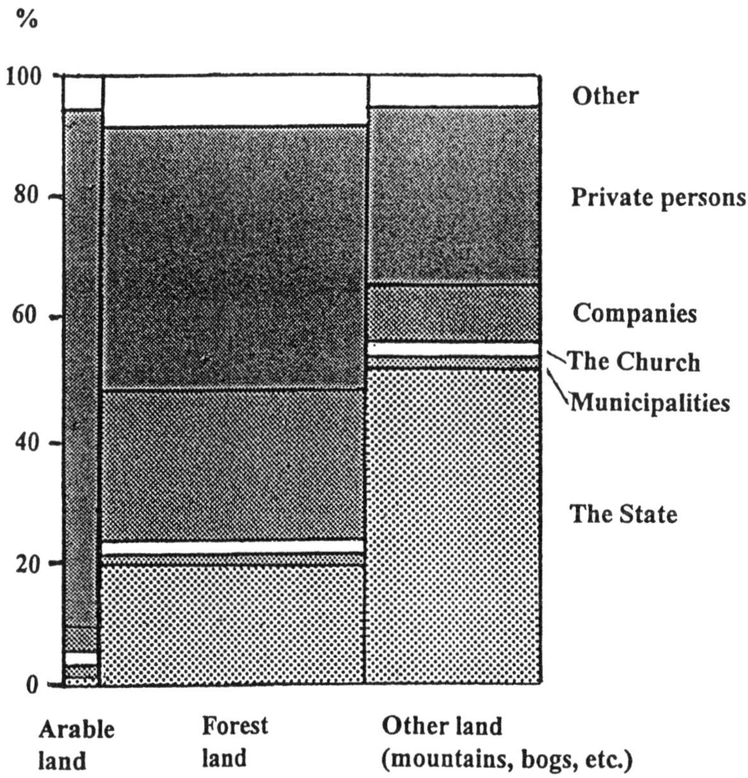

Figure 5.1 Ownership of arable, forest and other land in the 1980s; column widths are proportional to areas (*Source:* SCB 1993h).

ket, whereas the remainder changed hands as a result of in-family purchase, gift or inheritance. Thus, about 4000 purchases took place every year on the open market (SCB 1992b) and, assuming the average acreage of these properties to be 40ha (i.e. slightly above the overall average), the annual turnover on the open market must have been about 200000ha, that is 0.5% of Sweden's rural areas and about 1% of its productive agricultural and forestry acreage.

Data taken from open-market transactions for predominantly agricultural and forestry undertakings show that values diminished in real terms until 1987, when the rules of acquisition were liberalized and the market virtually deregulated. Until then, purchases were subject to price assessment and the purchase price was not allowed to differ appreciably from the yield value of the property. The prices of mainly forestry undertakings rose steeply between 1987 and 1990, after which they fell rapidly. This movement of values was probably attributable to cyclical fluctuations and the concomitant changes in timber prices. No such steep rise and recession can be observed in the case of predominantly agricultural undertakings.

The average price of agricultural and forestry properties sold on the open market in 1991 was about SEK900000, buildings included (SCB 1993g).

The following account of the hectare values of agricultural and forestry land in 1992 is based on LMV (the Central Office of the National Land Survey) material (1993d). The average value of agricultural land varies a great deal but ranges between SEK5000 and SEK15000 per ha in the south of Sweden and SEK4000 and SEK6000 in the north. Hectare prices, however, especially in the south of Sweden, vary a great deal depending on the yield capacity of the land: in forest areas they can be just a few thousand SEK, whereas the most fertile soils in the extreme south can cost anything up to about SEK40000.

Forest land values depend partly on productive conditions, but the existing volume of timber is the factor by which values are influenced most. Timber prices are to a great extent governed by the world market, so that forest land prices fluctuate with the international economy. The market value of timber in the autumn of 1992 was about SEK120 per forest m^3; this corresponds to a land value of about SEK15000 per ha for semi-mature forest. Values vary from one region to another, and in the south of Sweden in recent years they have been 75–100% higher than in the northernmost parts of the country. The hectare values of semi-mature forest in 1992 were about SEK20000 in the south and 10000 in the north. Areas that are ready for felling fetch roughly twice as much.

5.2 Undeveloped building plots

There are extremely few regions with any true shortage of undeveloped or raw land for urban development, because so little of Sweden has been built on. The planning processes, however, are often protracted, with the result that undeveloped land is not always available when needed. Existing undeveloped sites, on the other hand, are normally available for instant building development, the necessary planning formalities having been completed at the parcelling stage. The term plot (*tomt*) as used here refers to an undeveloped property (*obebyggd*) with a specified development purpose (*bebyggelseändamål*).

The number of plots is relatively large, and most of them are intended for single-family and secondary housing development (Table 5.1). The plot statistics refer to Sweden as a whole and make no distinction between urban and rural areas. The breakdown of single-family housing plots between these types of area is unknown. Plots for rental housing and industry, on the other hand, are mostly located within urban areas, and plots for secondary housing outside them.

Table 5.1 Undeveloped plots, 1992. Number of plots, number of plots in relation to total number of developed properties, percentage turnover of plots and increment of developed properties during the year. *Sources:* SCB (1993e, 1993i; own recalculations).

Plots for	No. of plots	Plot reserve (%)	Turnover (%) Deeds	Turnover (%) Straight purchase	Increment of developed properties
One- and two-dwelling buildings	62100	4	9	3	16400
Secondary homes	98900	17	3	1	2700
Tenement buildings	4100	4	12	2	1300
Industrial properties	12000	23	6	1	800

The plot reserve (meaning the number of plots in relation to developed properties) is strikingly large for industry (23%) and secondary housing (17%) development, but a good deal smaller for permanent housing, i.e. single-family and rental dwellings, non-housing premises included (4%; Table 5.1). Comparing the reserve in 1992 with building development that year (the increment of developed properties for the year), one finds that, theoretically speaking, the reserve is sufficient for a good many years to come, especially as forecasts indicate a low rate of building development for some years to come. The high level of reserves in relation to actual building development may be partly attributable to the plots having been parcelled in years with quite different expectations.

It is not even certain that all the plots prepared will be used for their intended purpose. Developments may have eliminated demand for them because they are in the wrong locations. Or the owners may be hanging on to them for one reason or another. The latter is probably a common occurrence with plots for secondary housing development, which may be intended for the use of future generations or may be a kind of nest-egg for the owners. The number of plots in relation to the number of second homes in Sweden, as well as turnover, tends to bear this out.

A rough picture of the regional distribution of the reserve of building plots in 1992 can be obtained by considering the reserve as a percentage of developed properties. This overview is based on county-by-county analyses of SCB (1993i).

Permanent housing purposes (single-family) In most counties the reserve was between 3% and 6%. Reserves were low in the big-city counties (3%), and relatively high in the three northernmost counties (6–7%). Otherwise, it is hard to discern any geographical patterns.

Secondary housing purposes Reserves varied between 10% and 30%. They were smallest in the big-city and adjoining counties (10–15%), largest in the five Norrland counties and in the so-called "forest counties" of Värmland and Kopparberg bordering on Norway (20–30%). Other counties in the south of Sweden normally had reserves of about 15–20%.

Rental housing purposes County reserves varied between 3% and 6% but were mostly around 4%. No regional pattern was really apparent, except that the reserves were smallest in the counties around Stockholm.

Industrial purposes Reserves varied a great deal from one county to another (15–30%), but it is hard to find any regional pattern, except that undeveloped industrial land was in plentiful supply in all five Norrland counties. Reserves here varied between 24% and 30%. In other words, for every third or fourth existing industrial facility, one more could be added more or less immediately. This kind of expansion is hardly reasonable, and instead the plots have been arranged against the contingency of sudden demand. The County of Stockholm had a large reserve (30%), even though the region is shedding industry.

Building plot turnover in 1992 was roughly half what it had been a year earlier. Turnover in relation to the total number of plots was between 3% and 12% (Table 5.1). It was highest for rental housing plots (12%), followed by single-family housing plots (9%), industrial plots (6%) and secondary housing plots (3%). Probably this is symptomatic of planning and building being more in step in the housing sector than in the case of industrial start-ups. Industrial land is often prepared without any particular interested party or use in mind. It is also worth noticing that straight purchases of building land make up only a small proportion of all changes of ownership, especially where rental housing and industrial plots are concerned (Table 5.1). This means that most changes of ownership involve some kind of agreement-conditioning community of interest between buyer and seller. Otherwise, plots are often prepared in connection with being built on, in which case they never change hands in the unbuilt state.

Price movements after 1981 will also be described with the aid of purchase price coefficients for permanent and secondary housing plots (Fig. 5.2), in SEK per m^2 building rights for rental housing plots and SEK per m^2 building plot for industrial development (Fig.5.3). Figures refer to both current and fixed prices, but in the main body of the text only fixed price movements will be remarked on.

The change in the purchase price coefficients for single-family and secondary housing plots shows that real prices fell until 1986, and that this was followed by a resurgence, especially for permanent housing plots (Fig. 5.2). The changes occurring in the 1980s coincided on the whole with changes in housing production, although the steep decline in single-family housing production during the first five years of the 1980s led to a more marginal drop in the prices of building

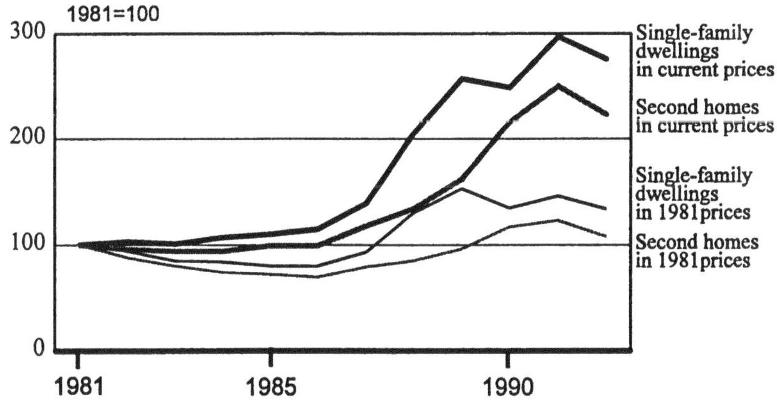

Figure 5.2 Movements of prices, 1981–92, of plots intended for permanent and secondary housing (purchase price coefficients) (*Source:* SCB 1993e; own recalculations).

plots (see Fig. 2.1). Production of single-family housing fell by 60% between 1981 and 1986, while plot prices dropped by 20% in fixed price terms. This price movement, however, did not tally with the cyclical change in GDP terms (see Fig. 1.2). Instead, the hot years in the capital, property and share markets at the end of the 1980s, like the ensuing downturn (see Ch. 1.2), are reflected in both housing production and plot prices. The movement of secondary housing plot prices describes the same pattern as for permanent housing plots.

The upturn in prices of industrial land (Fig. 5.3) came in 1984, roughly at the time when industry began increasing its investments in machinery, buildings and plant (see Table 1.10). Investments peaked in 1989. Thus, the fall in building plot prices began a year ahead of the fall in building and plant investments, which

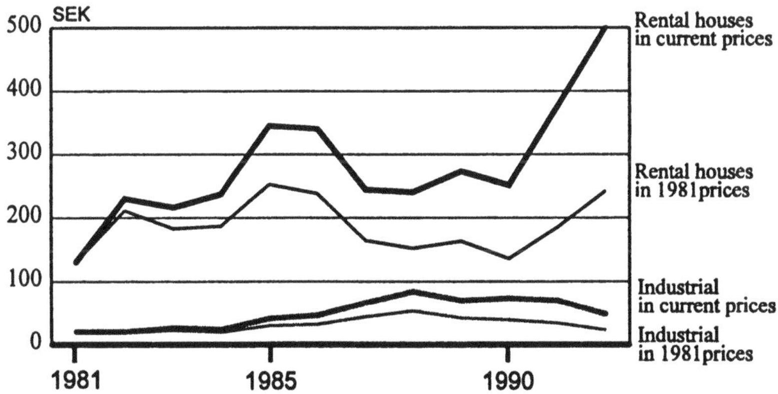

Figure 5.3 Movements of prices, 1981–92, of plots intended for rental housing (SEK per m^2 building rights) and industrial (SEK per m^2 plot area) development (*Source:* SCB 1993e; partly own recalculations).

perhaps is not very surprising, considering that land purchases precede building. Not too much should be made of these statistics, however, because the input material is limited (about 150 purchases a year until 1985, after which the number rises to just over 400 a year between 1988 and 1990, dropping back to just over 100 in 1992).

It is difficult, not to say impossible, to put an interpretation on the movement of rental housing plot prices (Fig. 5.3). The statistics are disrupted by the small number of straight purchases (varying between 30 and 100 a year) and by uncertainty as to whether the plots were used for housing or non-housing development. Besides, straight purchases account for only one sixth of the market turnover. This small proportion of straight purchases suggests that dealings in rental housing plots often form part of a contract package relating to more than just the land, so that the free market is limited. Nor is it uncommon for rental housing plots to be sold as a company, in which case they are not registered as real-property conveyances.

The prices of plots sold on purely market-related conditions fluctuate wildly all over the country, as can be instanced by means of average selling prices in 1992 (SCB 1993e).

Permanent housing purposes (single-family) The average price of a plot was just over SEK400000 in the County of Stockholm and upwards of SEK300000 in the counties of Göteborg and Malmöhus, whereas in other counties, with one or two exceptions, it varied between SEK60000 and 180000. The County of Halland, south of Göteborg, was one of the exceptions, with prices averaging not quite SEK300000.

Secondary housing purposes Inter-county price variations described more or less the same pattern as for permanent housing plots. Prices were highest in the County of Stockholm (SEK300000), followed by the counties of Göteborg, Halland and Malmöhus (about SEK200000). Elsewhere in southern Sweden, prices averaged about SEK100000, and in the forest regions of western and northern Sweden they ranged between SEK50000 and 75000. Plots with water frontages cost a good deal more, occasionally several times more, than others. A rough county-by-county comparison of price relations between secondary and permanent housing plots suggests that a secondary housing plot fetches about three quarters of the price of a permanent housing plot in the same county.

Rental housing and industrial purposes The official statistics cannot be meaningfully broken down to regional level. Generally speaking, the price of land rises with the size and centrality of a location. This is especially true of rental housing with non-housing facilities. In 1992, land for rental housing purposes was fetching SEK500 per m^2 building rights, and industrial land not quite SEK50 per m^2 building plot (median figures for the country as a whole).

5.3 Site leasehold

Site leasehold (*tomträtt*) is a relatively common form of tenure in certain municipalities, and therefore requires a word of explanation. Site leaseholds for housing purposes run for 60 years, after which the owner (almost invariably a municipality) is entitled to cancel the contract every 40 years. The land can then be conveyed to a new party for a different kind of activity offering a better return. Non-housing site leaseholds come up for renewal every 20 years. If the municipality requests termination of the site leasehold, lessees receive compensation for their investment. The lessee has no right of cancellation.

The ground rent (*tomträttsavgäld*) is set at the first conveyance and is usually adjusted at ten-year intervals. Normally it has to reflect a real rate of land interest on the market value at the time of adjustment. The rate of interest is 3.75% for index-linked land with a ten-year ground rent period. In municipalities where real values are falling, the ground rent interest may be 5–6%, and in certain first-time conveyances it can be anything up to 10% with ten-year ground rent periods. The methods used for determining the other ground rent factor, i.e. market value, vary considerably from one municipality to another. In localities where site leasehold is widespread, moreover, new ground rents can be difficult to set, because a lack of comparable property purchases makes it hard to determine the market value of the land. If the parties fail to agree on the new ground rent, the matter can be taken to court.

Site leasehold occurs on a considerable scale in about 50 municipalities and on a small scale in another 50. The number of site leaseholds at the latest more detailed count, at the end of the 1970s, was about 90000, which is barely 5% of the national property stock. Of these, 80000 were for single-family dwellings, 7000 for multi-family dwellings and the remainder mostly for commercial properties. One third of all site leaseholds existed in the Greater Stockholm region (SOU 1990). The 1980s brought a steep decline in the use of site leasehold for new building development, and in 1990 only 2% of residential estates and 10% of apartments were conveyed on this basis. The corresponding figures ten years earlier were much higher, namely 15% and 25% respectively (SCB 1992a). About 15000 new site leaseholds have been created since the count was taken in the 1970s (CFD 1993).

Half (29000) of all properties within the boundaries of the municipality of Stockholm are conveyed by site leasehold. The municipality of Västerås has also made very extensive use of site leasehold, about 40% (10000) of all properties being conveyed on this basis. Other municipalities with large numbers of site leaseholds (3000 or more) are Göteborg (11000), Malmö (5100), Norrköping (3900), Luleå (3500), Gävle (3200) and Karlstad (3200). In all of these communities, site leaseholds account for 15–20% of the entire property stock (calculations of CFD 1993).

Stockholm, then, is Sweden's largest user of site leaseholds, with ground rent revenues exceeding SEK1000 million in 1991 (USK 1992).

5.4 Single-family dwellings

Sweden has upwards of 4 million dwellings. These break down into 1.9 million single-family dwellings and 2.2 dwellings of other kinds, most of them multi-family.

Of the single-family dwellings, 1.3 million are in urban areas, and of these 70% are detached houses and 25% are terraced, semi-detached and linked houses (Table 5.2). Of these dwellings, 90% are owned by private individuals and the remaining 10% by non-profit housing utilities or tenant-owner co-operatives. One twentieth of the privately owned houses are let. More than half of all single-family dwellings in urban areas were built after 1965, i.e. after the commencement of the Million Homes Programme, and 170000 date from the 1980s.

Table 5.2 Categories of single-family dwellings, 1990, by urban and rural areas *Source:* SCB (1992a).

Category	Urban	Rural
Detached	907000	506000
Detached for two families	80000	34000
Semi-detached, etc.	338000	8000
Total	1325000	548000

There are 550000 single-family dwellings in rural areas, and 160000 of these are farmhouses. More than 90% of single-family dwellings in rural areas are detached houses, and most of the remainder are two-family homes. These houses are nearly 100% privately owned. One home in ten is let. Single-family dwellings in rural areas are often fairly old, half of them having been built before 1940, 50000 were built after 1980 (SCB 1992d).

With the exception of the population and housing censuses, official housing statistics normally make no distinction between homes in urban and rural areas, hence the simultaneous treatment of both types of area in the following description of the single-family housing market. There is in fact some justification for lumping urban and rural properties together, because a large proportion of homes in the country are inhabited by people working in towns. The statistics for the single-family housing market come from SCB (1993a) unless otherwise indicated.

Up until the beginning of the 1980s, single-family dwellings were produced mainly as detached houses. Just over a quarter were terraced or linked houses. The latter figure has increased since then and in 1990 accounted for half the new homes produced. The same period has also seen a notable change in the tenure of new homes. Whereas virtually all new homes used to be owned freehold, this form of tenure has to a great extent been superseded by tenant-ownership and rental tenure. Freehold ownership of new single-family dwellings has now fallen below 40%. The change can be partly put down to government credit subsidies favouring tenant-ownership and rental tenure rather than freehold tenure.

Since the 1960s, the average spatial standard for the single-family housing stock as a whole has generally risen from three rooms to upwards of four. There are, however, tenure-related differences. During the 1980s, new rented homes had three rooms, tenant-owner homes four and freehold homes five. Kitchens and ancillary spaces are always added. Differences in room size were also reflected in the floor spaces. This can be illustrated by means of detached and semi-detached housing estates. Rented homes averaged $80\,m^2$, tenant-owner homes $100\,m^2$ and freehold $120\,m^2$.

Modern building methods can be described in terms of the selection of materials for production in the 1980s. This description applies only to residential estates, but it probably also holds good for other kinds of production. About 90% of the houses have timber frames, whereas the remainder have frames of concrete or other material. Two thirds have wooden façades and the rest have mainly stone cladding. There are great regional differences in this respect. In the extreme south of Sweden, only a quarter of the 1980s houses have wooden façades, as against nearly 100% in the more northerly parts of the country.

It was uncommon in the 1980s for single-family houses to be built on site. Instead, more than four out of every five houses were delivered prefabricated (STR 1993). Only one third of these factory-made houses, however, were entirely prefabricated, whereas the remainder presented various degrees of prefabrication.

The production costs of dwelling units in apartment blocks (the cost of the building) were shown in Figure 1.11. Detached and semi-detached residential estates cost 10–15% less to produce, because of the slightly diminishing marginal cost of extra spaces in dwellings of this kind. On the other hand, costs over and above the cost of production, such as connection to water and sewerage systems and streets, as well as land costs, are normally higher per m^2 building area. Consequently the total cost per m^2 of single-family dwellings is almost on a level with dwellings in multi-family buildings. In 1991 the average production cost of a single-family dwelling was SEK10500 per m^2 building area, and this included building costs of SEK8200.

At the beginning of the 1980s, the annual turnover in single-family dwellings on the open market was not quite 45000 properties. This gradually increased, reaching 60000 units in 1988. The increase was probably connected with a rise in real earnings and with the deregulation of the credit market. Turnover remained high up to and including 1991, but then in the following year it plummeted to 33000 properties, a fall of 40% (SCB 1993e). This low level of turnover persisted during 1993.

An annual turnover of between 45000 and 60000 properties in the open market corresponds to a turnover of 3–4% of the total stock. The number of conveyances with a price-conditioning relationship between the parties (in-family transactions, gifts, inheritance, etc.), is unknown but probably amounts to a few more per cent. Average tenure of a single-family dwelling can thus be estimated at about 20 years.

The number of sales, like house prices, fluctuates during the year. Normally more single-family dwellings are sold in summer than in winter: sales in June, for example, are two to three times as high as in February. Normally, prices go up during the spring, peak during the summer, then in September begin to come down again, bottoming out in December. Another price peak comes in January, and is presumably connected with the rules of capital gains taxation (SCB 1993n).

The rise in house prices was marginal in the extreme during the 1980s and all the more noticeable towards the end of the decade. The movement in real prices was, if anything, *negative* at the beginning of the decade, with prices in real terms continuing to fall until 1985–6. During the next four years, prices rose by 40%, reaching a peak in 1990. By 1992 prices were back to about the same real level as at the beginning of the 1980s, and one year after that they were lower still. Thus, in real terms prices had reverted to the state of affairs preceding the property boom. In nominal terms, however, prices did not begin falling until 1992, at the same time that turnover declined rapidly.

General economic developments probably contributed towards the rise in prices during the late 1980s, just as they helped to bring about the fall after 1990. We find that the real interest rate (interest minus inflation) was low at the end of the 1980s but increased slightly in 1991 and, above all, in 1992. This, coupled with a tax redeployment that reduced tax reductions for mortgage interest payments, made mortgages a great deal more expensive for households. There was also a rise in running costs. During 1993, however, interest rates fell steeply, partly because of the international movement of interest rates, which had the effect of reducing interest outgoings on a given amount of borrowing. This ought logically to raise house prices in the long term, at least when households come to take a more optimistic view of the economic future. One sign of this possibly being the case is the stability of prices all through 1993 (LMV 1994a). Another upward force in the long term may be that living in a house of one's own has come to be cheaper compared with apartment rents (LMV 1993d).

The regional movement of prices can be illustrated by means of the trend in purchase price coefficients in the counties of Stockholm and Upper Norrland (Fig. 5.4). In real terms, prices in both regions fell by 20% in the period ending in 1986. After this they rose, only to fall again after 1990. The price rise in the Stockholm region was especially noticeable (more than 50% in three years), as was the ensuing decline. In Norrland, prices at the end of the 1980s rose to the 1981 level only, and the ensuing decline was fairly limited. Developments in the Göteborg and Malmö regions almost mirrored those in Stockholm, whereas price movements elsewhere in the country tallied with those in Upper Norrland. Thus, the big-city housing market was dragged along by the general price rise in a way that had no counterpart elsewhere in the housing market. This finding, however, must be treated with some caution, because the statistical material takes account of only the regional markets. Rural sales may have had a restraining effect on price trends outside the large cities. Special surveys are needed, however, to settle this point. Between 1992 and 1993, the fall in prices appears to have continued in all submarkets, matching the decline for the housing stock generally (see Fig. 5.4).

The average price of a single-family dwelling in 1992 was nearly SEK0.7 mil-

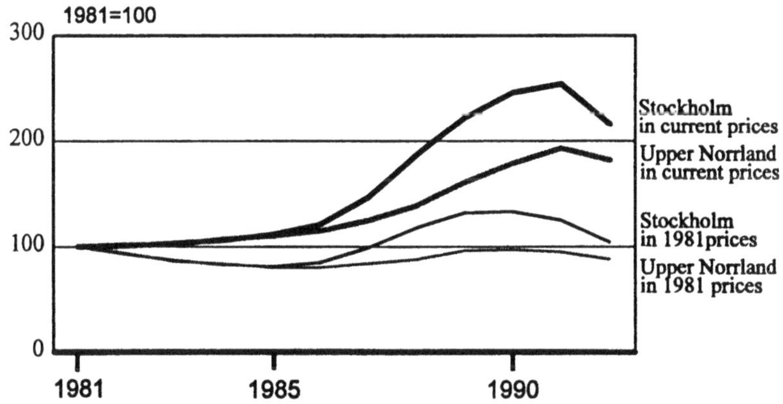

Figure 5.4 House price movements in Greater Stockholm and Upper Norrland, 1981–92 (purchase price coefficients) (*Source:* SCB 1993e; partly own recalculations).

lion, but prices varied according to location, type of house and age. That same year, the average purchase price of a detached house in the Stockholm region was SEK1.3 million, and in the Göteborg region SEK0.9 million. In the counties immediately surrounding Stockholm and in the Malmö region, prices were somewhat lower, at roughly SEK0.7 million. Elsewhere in the country, the normal price as a rule was about SEK0.5 million. Especially low prices occurred in rural communities at a great distance from larger localities. Here prices averaged SEK0.3 million or sometimes even less. In addition to varying between different parts of the country, prices can also vary locally within a single region. In the Stockholm municipalities of Danderyd and Lidingö, for example, houses were fetching just about twice as much as in several other municipalities in the same region (SCB 1993e). The prices of terraced and linked houses are the same, on the whole, as those of detached houses if the buildings are of comparable area and age.

As a benchmark for house prices, we can say that they averaged just over SEK5000 per m^2 dwelling space in 1992 (SCB 1993e). This does not allow for regional variations. Prices were a good deal higher mainly in the Stockholm and Göteborg regions, and a good deal lower in rural areas.

The movement of housing costs for freehold homes more or less matched inflation until 1990, after which there was a distinct rise, especially in running costs. The high level of prices at the end of the 1980s also involved some families in high capital costs.

Capital costs (interest expenses and instalments) normally account for the greater part of housing costs in a single-family dwelling. When a house changes hands, capital costs often account for more than 80% of the total housing costs, with running costs making up the remainder. In cases where the house has been owned for some years, a better situation normally applies, in the sense of capital costs being a good deal lower because of the effects of inflation and, to some extent, to previous amortization payments. After some years, capital costs then average two thirds of total housing expenditure (SCB 1993a).

Rising costs and perhaps too the difficult employment situation have led to a rise in the number of repossession house sales as owners fail to keep up their payments. The number of repossession sales, however, equals only 1% or so of the total market turnover. It is not uncommon, though, for properties to be sold in the course of the repossession process, instead of being put up for auction at the end of it, and so the number of property owners with acute financial problems may be greater than repossession sales suggest. Calculations to indicate the scale of negative equity suggest that up to 5% of single-family dwellings are mortgaged above their market value (Råckle 1993). This was made possible by the movement of prices in the 1980s.

5.5 Second homes

Figures for the number of second homes in Sweden vary, owing to different statistical sources employing different definitions. National estimates, however, point to a total of 670000 houses (SCB 1993h). Secondary housing development concentrates on areas outside the large cities and along the coasts. Figure 5.5 shows the regional breakdown. The map does not show it, but within each county the heaviest concentration occurs in coastal areas and alongside lakes and watercourses. Normally there are also heavy concentrations in mountain regions near winter sports centres (SOU 1979).

Secondary housing plots vary a good deal in size, but the commonest size is between 1500 and 2000 m^2. Most second homes are built on site. Half of them are self-built by the owner. Such homes are normally intended for summer use, but many of these houses are "winterized" for year-round use, i.e. built to appropriate heating and insulation standards (Larsson 1979). The average dwelling space is 50 m^2. Supplementary outbuildings are also common (SCB 1991b).

The construction of second homes (as of permanent homes) gathered speed after the Second World War, culminating in the 1960s and 1970s (Fig. 5.6). During the 1980s, by contrast, output was relatively moderate, and during the boom years following 1985 only 20000 houses were built, which makes an average of 4000 a year. During the recession year of 1992, output nose-dived to 1000 units (STR 1993). It should be added that most of the second homes built before 1930 were originally permanent homes.

About half the inhabitants of Sweden have the use of a second home. Ownership is commonest, but almost the same number have the use of a second home through relatives. Not quite 10% of the population rent their second homes (SCB 1993a). Nearly all second homes in Sweden are owned by private individuals or estate trustees (SCB 1991b).

Turnover in second homes on the open market is small in relation to the stock. Since 1980 about 10000–15000 properties have changed hands every year,

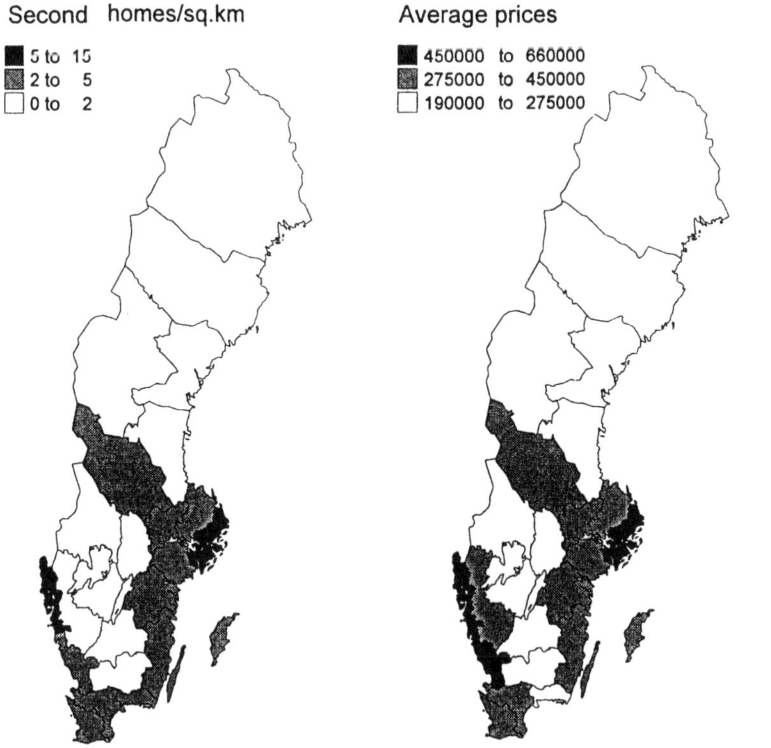

Figure 5.5 Number of second homes per km² and average prices in 1992 (by counties) (*Sources:* SCB 1993e, SCB *Statistical yearbook* 1994).

which is about 2% of the total stock. The number of second homes passing into new ownership is more than twice this figure. In other words, a second home is more likely to change hands by being bought and sold within the family or through gift, inheritance, and so on, than by being put on the market (SCB 1993e). In 1992 the number of purchases on the open market fell by 40%.

In nominal figures, second home prices rose throughout the 1980s. In real terms, however, they fell by about 15% between 1981 and 1986, rising thereafter by 50%. Prices did not fall again until 1992, and this fall continued in 1993. As a result, real prices reverted almost to their level at the beginning of the property boom. Regional price movements follow the same pattern, as instanced in Figure 5.7 with figures for the Stockholm region and the south of Sweden. Developments on the west coast of Sweden were much the same as around Stockholm, whereas developments elsewhere in the country were more similar to those illustrated for the south of Sweden. The interior of Norrland was the excep-

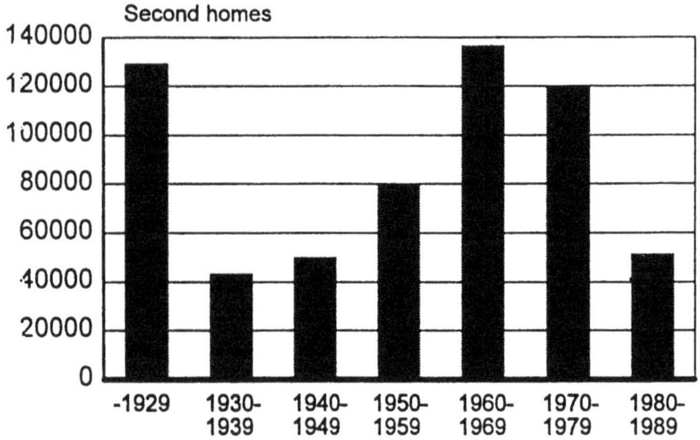

Figure 5.6 Number of second homes by age (thousands of tax assessment units) (*Source:* SCB 1991b).

tion to this rule. Prices there fell by 20% until 1985–6, after which they rose only insignificantly until 1991, before dropping still further (SCB 1993e). These houses in sparsely populated regions appear to be the least attractive.

The average price of a second home in 1992 was SEK350000, but this conceals large regional differences. Moreover, waterfront plots invariably fetch far higher prices than plots without any water contact. Prices are highest in the most attractive and therefore densest areas. The Stockholm region has the highest average prices (about SEK650000 on average in 1992) and large parts of the interior and Norrland have the lowest prices (about SEK200000). Houses in attractive mountain areas normally fetch high prices. Apart from demand, the need to build for winter conditions, and to a high standard in other respects, makes for expensive production in mountain areas.

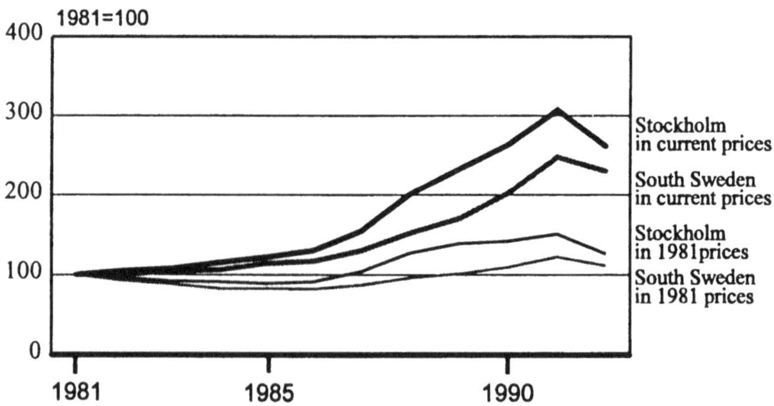

Figure 5.7 Movement of second home prices in the County of Stockholm and the south of Sweden, 1981–92 (purchase price coefficients) (*Source:* SCB 1993e).

5.6 Tenement buildings

Sections 5.1–5.5 have dealt with the market for properties that are normally managed under simple conditions, in so far as ownership and use are usually in the same hands, or else the entire property is leased or let. Forms of management for multi-family housing, non-housing rental properties and industrial facilities are often more complex. The properties may be partly or wholly let and, in that case, to one or more tenants. In addition, they are often put to mixed residential and non-residential uses. The difficulty involved in clearly distinguishing properties with different uses has resulted in rental apartment buildings and rental non-housing premises being grouped into a single category in Sweden's purchase price and tax assessment statistics, namely "rental properties". Industrial premises, on the other hand, are separately accounted for in the statistics.

The manner of statistical classification has left its mark on the account that will now be given of the rental property market. The section begins with a description of property stocks and property markets. Next comes an account of stocks of rental dwellings and non-housing properties, together with their rental levels. The description of industrial properties comes in Chapter 5.7. Tax-exempt properties, mainly comprising public sector buildings, are less often put on the open market, and are therefore dealt with only briefly in Chapter 5.8.

Property stocks

For the purposes of property tax assessment, rental properties (*hyreshus*) are buildings fitted out for the accommodation of at least three families or for use as offices, shops, hotels, restaurants or other commercial function.

Rental properties fall into three groups: those mainly consisting of housing, those comprising housing and non-housing premises, and those mainly comprising non-housing premises. The main criterion of classification is whether the dwelling space comprises more than 75%, between 75% and 25% or less than 25% of the total floor space of the property. Table 5.3 shows the number of units of each category in Sweden. Out of a total of 111000 tax assessment units, half are more or less pure housing complexes. About 25000 are mainly pure non-housing properties or service buildings. The table also shows the total assessed value of each property category. The market values are approximately 2.5 times higher.

A division of the assessed value of rental properties regionally by metropolitan counties (corresponding basically to the metropolitan regions) and other counties (Table 5.3) shows about 30% of the value to be located in the County of Stockholm and about 10% in Göteborg and Malmö respectively. Upwards of 50% therefore is located elsewhere in the country. These figures are worth comparing with the corresponding population percentages, which are about 20%, 10% and 10% respectively for the metropolitan counties and 60% for the other counties. This means that Sweden's real-property capital on the apartment side is concentrated within the metropolitan counties, and above all in the County of Stockholm. This is not very remarkable, considering that residents of large cities are relatively often flat-dwellers.

Table 5.3 Combined housing and non-housing rental properties; number of tax assessment units and assessed value, 1993; total and by metropolitan and other counties. Sources: SCB (1993i) and LMV (1994b); own recalculations.

Category	No. of units	Assessed value (SEK bn)	Distribution of value: Stockholm	Göteborg	Malmö	Rest of Sweden
Mainly housing	62600	230	28	12	9	51
Housing and non-housing	22100	80	41	9	9	41
Mainly non-housing	19500	110	52	11	5	32
Offices on industrial land	1400	5	29	28	9	34
Hotels and restaurants	2800	10	40	11	4	45
Kiosks	2500	1	19	11	5	65
Total	110900	435	37	11	8	44

On the other hand, the concentration of commercial real-estate capital in Stockholm is truly conspicuous. This is where we find about half of the country's rental buildings with non-housing premises – a state of affairs that is partly attributable to relatively high property values.

Of the total real-estate capital in Sweden's rental building sector, aggregating the values of both housing and non-residential, some 30% is owned by municipal non-profit housing utilities, 25% by Swedish limited companies, 20% by tenant-owner associations, nearly 15% by private individuals and upwards of 10% by other legal entities (SCB 1990b). Industrial premises and publicly owned premises are not included in these figures; see instead Chapter 5.7 and Chapter 5.8 below.

Prices Between 2500 and 3000 rental properties normally come on the open market every year (SCB 1993e). Price movements between 1981 and 1992 are shown in Figure 5.8, which refers to all types of rental property nationwide. Nominal prices rose uninterruptedly during the 1980s, but in real terms one finds

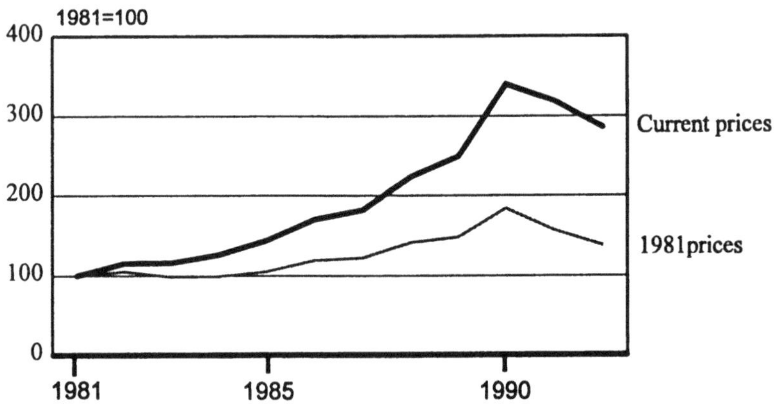

Figure 5.8 Movement of rental property prices, 1981–92 (purchase price coefficients) (*Source:* SCB 1993e; own recalculations).

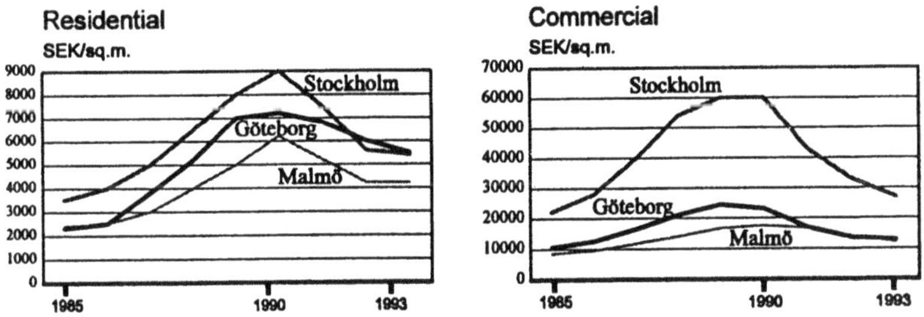

Figure 5.9 Movement of housing and commercial property prices in prime locations in Stockholm, Göteborg and Malmö, 1985–93 (current prices; SEK per m^2) (*Sources:* Catella 1994, SFF 1994).

that prices fell slightly at the beginning of the decade. From 1985 there was a more noticeable rise, which became really steep just before prices peaked in 1990, after which they fell sharply.

The great accumulation of real-estate capital in the large cities gives us cause to consider price movements there (Fig. 5.9). The price levels in our figure are admittedly approximations, but they show the changes quite clearly. The graphs refer to prime locations in large cities and are divided between residential and commercial properties. The prices are current prices for 1985–93 in SEK per m^2.

The average price level for prime-location housing properties in 1985 was about SEK3500 per m^2 in Stockholm and approximately SEK1000 less in the other two metropolitan cities. Prices then rose substantially, and in 1990 stood at SEK9000 per m^2 in central Stockholm and about SEK6000 in Malmö, with Göteborg coming in between. Comparing price movements in the prime locations of the metropolitan cities with the rest of the country, one finds that they were relatively strong in the metropolitan cities. After 1990, prices fell by one third in Stockholm and Malmö and slightly less in Göteborg.

Similar trends applied to commercial properties. In the Stockholm region, prices in nominal terms almost tripled between 1985 and 1990, in Göteborg they more than doubled and in Malmö they almost doubled. The price rise in Stockholm was notably steep compared with the rest of the country. In 1993, prices in the three metropolitan cities were back to about the same level as in 1985. On this subject we may also note that the prices in Figure 5.9 refer to prime locations and that in the rest of Stockholm prices during the period under consideration were barely half those that the figure indicates (Catella 1994).

The regional picture of falls in housing and non-residential property prices can be summed up in the light of LMV (1993d). Prices fell most steeply in the prime locations of large city regions. In central Stockholm between the summer of 1990 and the end of 1992, housing property prices fell by about 40% and non-residential prices by about 60%. The decline was not of the same magnitude in the rest of Stockholm, or in Göteborg or Malmö. It was lower still in the provincial county towns, namely 20% and 30% respectively for housing and non-housing properties, whereas in small municipalities without large urban communities it was 5% and 20% respectively. Price falls for combined housing and non-housing properties represented an average for the two strict categories. In addition, the fall in prices was about 15% greater for large properties than for smaller ones. Differences in the fall in prices probably reflect earlier differences in price rises, in the sense of falls being largest where the rises had been greatest.

Return on capital The Catella valuation company has in various connections presented estimates of the capital yields required from housing and non-residential properties in the metropolitan cities between 1985 and 1993 (Fig. 5.10). The yield expresses the ratio between market-based net operating income and paid prices. The reference is to prime locations. The return required fell from about 7–8% cent in 1985 to, at its lowest, 4–5% in 1989, rising thereafter to slightly higher levels than in 1985. Housing properties in central Stockholm deviated slightly, their yield requirement being roughly 1% lower for all years (Catella 1994, SFF 1994).

Thus, the infusions of capital received by the property sector at the end of the 1980s (see Ch. 1.2) led to a reduction in capital yield requirements. The acceptability of low yields, however, was probably attributable to the buyers' belief in continuing inflation and future rent increases, which together would raise the long-term yield in nominal terms. Probably there was also share speculation involved. When the economy deteriorated, the growth in lending decelerated, inflation was restrained and a glut occurred in the non-housing property market, this curtailed the upward spiral of prices. The final abolition of exchange regulations in 1989, moreover, made possible a growth of capital investment abroad. Non-housing premises were left untenanted and rents fell. Property companies developed liquidity problems, debts began to be foreclosed on, companies went bankrupt and the banking system was shaken to its foundations. However, property prices fell to such an extent that the capital yield increased.

In 1993 the government appointed a special group of experts on real-estate economics to take charge of property valuations when the State, in various ways,

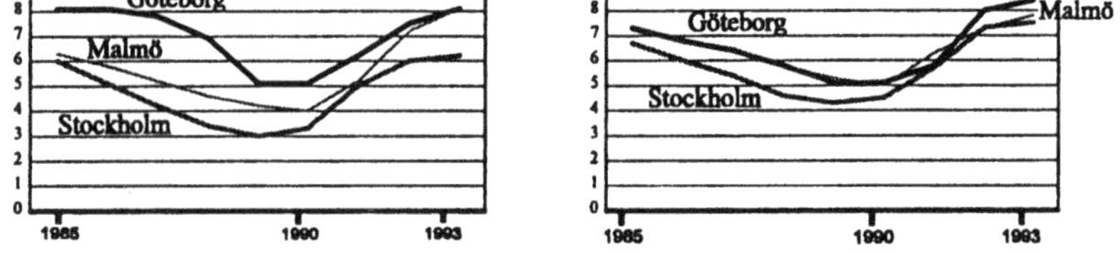

Figure 5.10 Development of capital yield, 1985–93, for prime locations in the metropolitan cities (housing and non-housing rental properties) (*Sources:* Catella 1994, SFF 1994).

helped to reconstruct the banking sector during the banking crisis. This group, which came to be known as the Valuation Board, analyzed the property market in order among other things to arrive at internal benchmarks for capital yield requirements in the valuation of properties. The analyses are presented in Bankstödsnämnden (1993), Bankstödsnämnden/Valuation Board (1994) and Valuation Board (1994).

The valuation experts estimated reasonable capital yield requirements against the background of current vacancy rates and subject to the assumption of a future real interest rate of 3% in the short term and 3.7% in the long term, together with 3.5% inflation. The results are summarized in Table 5.4, from which it can be seen that the capital yield requirement for housing is put at 7–12%, except for prime-location dwellings in the metropolitan cities, where a lower capital yield is prompted by expectations of higher rents being obtainable in future. Shops and offices come between 7% and 12%, whereas hotels and factories come between 10% and 16%. The total yield requirement in cash-flow analyses is about 2% higher than the direct capital yield requirement.

Table 5.4 Capital yield requirements for normal properties, 1994. *Source:* Valuation Board (1994).

Property	Capital yields (%)
Housing	
generally	7–11
conversion and new production	8–12
prime locations in metropolitan areas	5.5–8
Offices	7–12
Shops	7–12
Hotels	10–15
Industry	11–16

To augment the picture, we can take a regional description based on surveys conducted in 1992–3 by the journal *Fastighetsvärlden* (1993). The capital yield estimates are based on property purchases in about 40 large localities. Although the assessments rest on weak foundations, given that there were only a few purchases, certain cautious conclusions can nevertheless be drawn. Capital yield requirements are not really subject to any great regional differences. On the other hand, differences do emerge between prime and less-than-prime locations in a single locality where non-housing premises and shops are concerned. In prime or central locations, requirements are a couple of percentage units lower than in more peripheral locations. This means that the requirements are 9–10% for prime locations as against 10–12% for peripheral ones. The yield requirement for housing, usually 9–11%, is on the whole unaffected by location, the clearest exception to this general rule being Stockholm, where the yield requirements are lower.

Housing stock in rental buildings and tenant-owner associations

The following is an account of the dwelling stock in rental properties. Statistics, unless otherwise indicated, are taken from SCB (1993a).

There are 2.2 million dwellings in multi-family buildings. Of these, 40% of them are located in the three metropolitan regions. The remainder are also to be found in urban communities, mostly of the larger kind. Dwellings of this kind are few and far between in genuine rural areas.

The annual output of homes in multi-family buildings rose steadily from 30000 in 1950 to 75000 in 1970. After that the market became saturated and output from 1975 onwards was 15000–20000 dwellings annually. Production did not rise again until the end of the 1980s. After 1992 came a new decline, resulting from the recession and changes in the rules applying to housing loans (see also Ch. 2 and Fig. 2.1). As a result of a large output of new homes in recent times, more than half of all dwellings, i.e. 1.1 million, post-date 1960. Really old dwellings are rare: 1 in 5 dwellings were produced before the Second World War and only 1 in 20 before the First World War. Almost without exception, these older homes have been upgraded to modern standards.

Dwellings in multi-family buildings are more or less equally divided (one third each) between private interests, municipal non-profit housing utilities and tenant-owner associations. The State and county councils are marginal owners. Thus, two thirds of all multi-family dwellings are rental homes, whereas one third are included in tenant-owner associations, of which the tenants, by definition, are members. One fact worth noting is that private individuals are relatively numerous among owners of the older property stock, whereas the non-profit housing utilities have a relatively young property and, consequently, housing stock.

Spatial standards per person in dwellings are relatively high in Sweden by international comparison, and the same goes for the quality of apartments. Nearly half of all apartments comprise three or more rooms plus a kitchen. The stock of apartments is relatively young. On the other hand, Sweden being such a sparsely populated country, it may seem odd that apartments should make up such a large share (over 50%) of the total housing stock. The corresponding figures are a good deal lower in the Netherlands (about 30%) and the UK (20%). Conditions in Europe vary, however, and the situation in Germany is very similar to that in Sweden (Boverket 1993b).

Multi-family dwellings are usually constructed on site, even though the proportion of partly prefabricated dwellings has in recent years risen to a quarter of all new output. Roughly one dwelling in ten is entirely prefabricated. Production costs are shown in Chapter 1.4. Not quite a fifth of all production is undertaken by developers under their own auspices, the remainder being put out to contract. Building design has changed radically over the past 30 years, with three-storey buildings being superseded by two-storey ones. One reason for this is the introduction of rules requiring the installation of lifts in buildings of at least three storeys, which has made them uneconomical to build.

Rental apartments Rents in Sweden are usually paid by the month. Figure

5.11 shows the monthly rental cost in 1991 of a three-room apartment, by year of completion. As can be seen, rents vary according to the age of apartments – the rent for an apartment built at the end of the 1980s was about 30% more than that for an apartment dating from the 1950s. The rent per square metre is usually calculated on an annual basis, and for an average three-room apartment in 1991 it was SEK530.

Two years later the average rent for a three-room apartment climbed to almost SEK600 per m^2, corresponding to a monthly rent of not quite SEK4000 or an annual rent of SEK45000. Rents, it is true, have been clearly on the increase over a ten-year period, but, allowing for changes in the retail price index, the really large increases only came after 1991. The movement in rents was the same for apartments of other sizes. It is worth adding that rents in the large cities are 10–15% higher than elsewhere in the country.

Turnover of rental apartments or tenancy agreements is uncertain, for lack of official figures, but indirect estimates are possible. Housing surveys have shown that 25–30% of households renting their homes are tenants of up to two years' standing (Hägred 1994). This short period of residence for large groups suggests an annual turnover of about 15% of the apartment stock.

Tenant-owner homes First-time allocations of tenant-owner titles used to be mainly on a first come, first served basis. The slackening of the housing market in recent years, however, has more or less eliminated waiting lists. A deposit is payable for the first-time allocation, and this varies considerably from one part of the country to another. The average deposit in Stockholm and Göteborg, for example, was upwards of SEK200000 for a three-room apartment in 1991, whereas elsewhere in the country it was about SEK100000. Following this first-time allocation, tenant-owner titles are bought and sold in the open market. The

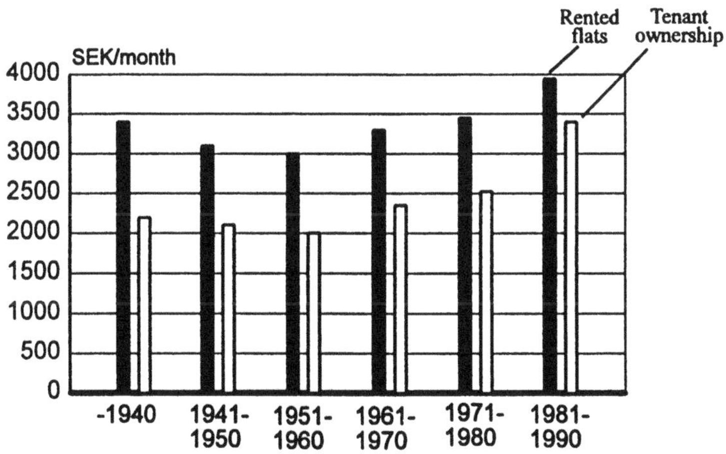

Figure 5.11 Average monthly cost of apartments with three rooms and a kitchen, 1991 (by age of apartment; rental and tenant-owner apartments) (*Source:* SCB 1993a).

purchaser pays for the right to join the tenant-owner association, and thereby acquires access to a home. The association, however, contracts the loans for the property, with the result that the purchase price does not include the mortgaged part of the property's value.

Figure 5.11 shows the average monthly charges for three-room tenant-owner apartments compared with rental apartments. The monthly charge includes a payment for interest and instalments on the association's loans, as well as running and maintenance costs. Tenant-owner apartments are consistently cheaper than rental apartments, but to some extent this is attributable to the members of the associations having paid a cash deposit. Moreover, the older the tenant-owner apartment, the lower the monthly charges will be. This is because the mortgage loans are often small ones. On the other hand, this makes older tenant-owner titles relatively expensive to buy; the lower monthly charges, in other words, have a boosting effect on selling prices.

Turnover of tenant-owner titles can be indirectly estimated on the same lines as for rental apartments. About 20% of tenant-owner households have been living in their apartments for up to two years (Hägred 1994), which suggests an annual turnover of about 10%.

Vacancies in the housing market, given a certain demand, are regulated by means of production and demolition. The latter, however, is unusual, and only about 2500 dwelling units annually have been eliminated in this way since 1980. Untenanted apartments and new production are, to some extent, inversely proportional to each other (Fig. 5.12). When vacancies decline, building sets in on a larger scale, and when vacancies rise, building production is cut back. As a result, both vacancy peaks and housing shortages are liable to be cyclically recurrent. A few years into the 1980s, there were 40000 untenanted apartments. The number fell gradually to 3500 in 1990, but by 1993 it was back to 40000, and in March 1994 had reached 55000 (SCB 1994c). This increase can also be put down to the economic situation and the rise in housing costs. In addition, demand is shifting in favour of smaller, relatively cheap homes.

Vacancy rates have a bearing on the movement of rents and, consequently, on the prices of rental properties. The rate of inflation and production costs are normally of secondary importance (Valuation Board 1994). This is above all true of non-housing rental properties, because pricing in this market is unregulated. But the same may come to be true of apartments, the rents for which are being increasingly adapted to market rents; the latter, at present, are considered to apply in large parts of Sweden, with the exception of town-centre apartments. In certain locations, landlords are even offering rent reductions and discounts as a means of attracting tenants. Thus, what used to be a thoroughly regulated housing rental market is gradually turning into a market with relatively free pricing. In the long term this could lead to substantial rises in rents for town-centre apartments.

By the end of 1993, an estimated 70–80% of rental apartments had attained market rents. On the other hand, this had not happened in attractive locations, above all in central Stockholm and Göteborg, owing to rent control systems. There is, however, a strong expectation that central locations will have an impact

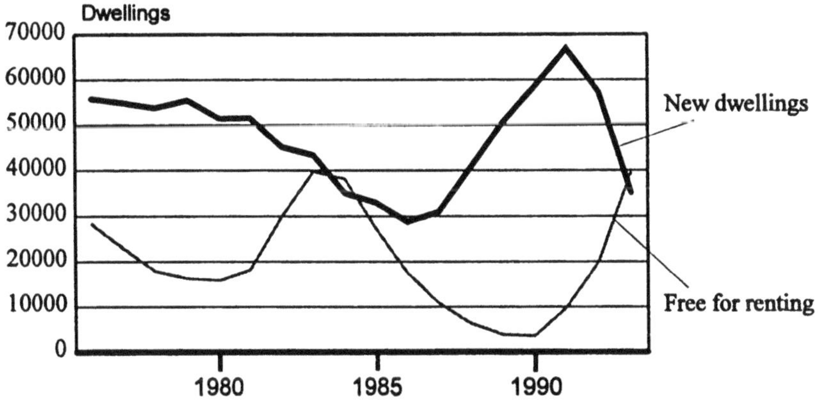

Figure 5.12 Untenanted apartments, March 1977 to March 1993, and new output for various years (*Sources:* SCB 1993a, 1994c; see also Fig. 2.1).

on rents (LMV 1993d). The relatively low requirements for capital yields, above all in downtown Stockholm, are probably an indication of this (see Table 5.4).

Vacancy rates in the housing market as a whole, however, are relatively moderate (2–3% of the apartment stock), although locally speaking they can be considerable. Moreover, the low level of new output planned for the next few years may lead to a drastic reduction in the number of untenanted apartments when the economy picks up again (LMV 1993d). Thus, there is liable to be a housing shortage later on.

Non-housing property stocks

Sweden has about 150 million m^2 of premises for the production of services. The four largest applications are education, offices, shops and caring services, in that order (Fig. 5.13), which between them occupy about 85 million m^2 or 55% of all non-housing premises. The remainder is used for wholesale trade, warehousing, hotels, restaurants, meeting places, workshops, day nurseries, sports facilities, banking and insurance, and accommodation.

Figure 5.13 also shows service sector floor space by ownership. For this purpose, the ownership categories employed are private, public and others. The term "private" includes private individuals, private undertakings, insurance companies and similar property-managing institutions. Public owners are the State, county councils, municipalities and municipal housing companies. Other owners include tenant-owner associations, co-operative enterprises and non-governmental organizations.

Over 25 million m^2 of non-housing premises are used for education, and these are almost entirely owned by public agencies, either directly or through publicly owned companies. About 20 million m^2 are used for office activities, ownership being more or less equally divided between the private and public sectors. This figure does not include office premises used as part of other activities mentioned in the figure (15 million m^2), nor does it include office facilities in conjunction

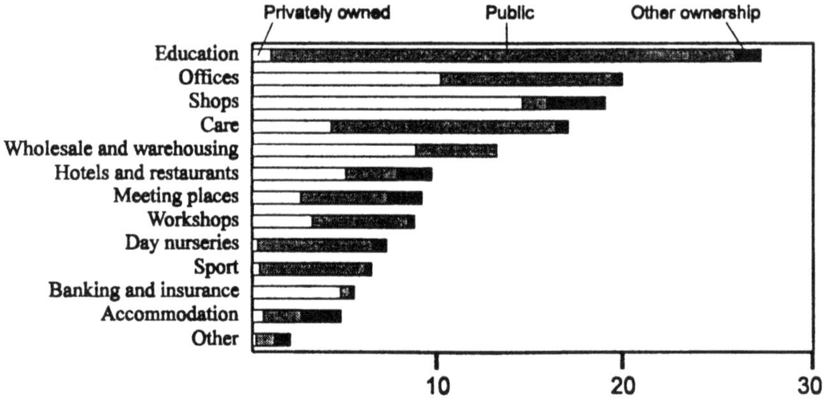

Figure 5.13 Service sector floorspace, by different types of premises and ownership categories (*Source:* Zingmark 1994).

with industrial activities (15 million m^2). Office floor space in Sweden, then, is a good deal greater than the figure suggests, namely 50 million m^2.

The third-largest use made of non-housing premises is for retail trade (nearly 20 million m^2). Food stores use roughly one third of this total. The private sector owns about 15 million m^2, whereas the co-operative and public sectors own the remainder. Caring premises occupy more than 15 million m^2 and are 75% owned by the public sector.

Roughly half the stock of non-residential premises has been built since 1970, which means that annual production since then has averaged 3.5 million m^2. The private sector has built for its own needs and for letting, whereas the public sector has built exclusively for its own needs.

Zingmark (1994) has collated figures concerning the letting of the stock of non-residential property. Between 50% and 75% of office and shop premises are let in the open market, but only 10–20% of the other categories of service premises. The reason for the low letting rate for other premises is that they are mainly owned and used by public institutions and non-governmental organizations. Added to this, the public sector is one of the large tenants, renting between 12 and 15 million m^2. It is also worth mentioning in this connection that roughly a quarter of the total industrial floor space of not quite 100 million m^2 is occupied on a rental basis.

Non-housing rents rose steeply at the end of the 1980s in current price terms, after which they fell by 30–40%. This development was distinct in the metropolitan cities, but the same trends were also at work in other parts of Sweden.

Non-housing rents in prime locations in 1993 were about SEK2000 per m^2 in Stockholm (Fig. 5.14) and about SEK500 less in other town-centre locations. Shop premises in corresponding locations can cost more than twice as much. Non-housing rents in less central parts of Stockholm and in other parts of Sweden, Göteborg and Malmö included, were about half those being charged in central Stockholm locations (Catella 1994).

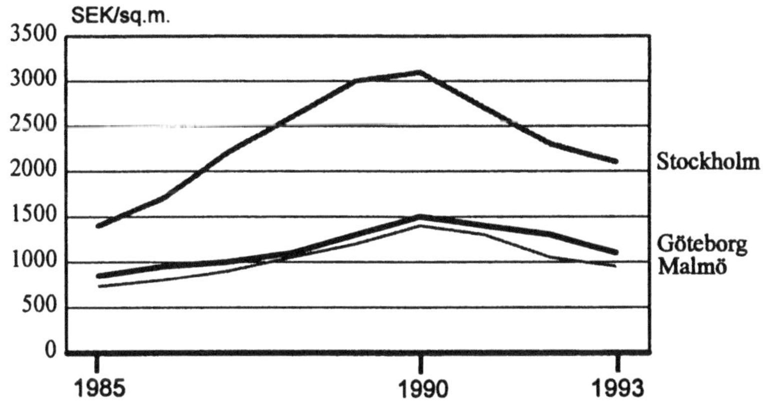

Figure 5.14 Non-housing rents in prime locations in Stockholm, Göteborg and Malmö, 1985–93 (SEK per m^2; current prices) (*Sources:* Catella 1994, SFF 1994).

Vacancies Great changes have occurred in supply–demand relations in recent years, in both the non-housing and property markets. There are several reasons for this. Office personnel strength has been reduced, at the same time as companies have been put out of business. At the same time, floor space consumption per employee appears to be diminishing. The readjustment entailed by reorganization measures in the public sector is leading to the cancellation of tenancy agreements with private property owners. At the same time, the public sector's own surplus is being put on the non-housing and property markets through publicly owned property companies (Zingmark 1994). In addition, production of non-residential premises ran at a high rate at the end of the 1980s. This had the effect of putting both rents and property prices under pressure, at least during a transitional period until low production results in vacant premises being fully tenanted once again.

Non-residential vacancy rates in 1992 were about 10–15% in the metropolitan cities and anything up to 15–20% in central Stockholm. In smaller localities the picture was more varied, but in some places vacancy rates could attain the same levels as in the metropolitan cities. Most localities, however, experienced vacancy rates of about 10% (LMV 1993d, Valuation Board 1994). It can take time for vacant premises to become occupied again with activities of different kinds owing to structural changes in the enterprise sector, and some observers are inclined to believe that vacancy rates will approach 15–20% in the next few years (Zingmark 1994).

Trade is one of the large demanders of premises and often pays the highest rents per square metre. The rented floor space is 10–15 million m^2. The boundary line here between retail and wholesale trade is uncertain. Assessments of the future indicate that great changes may be imminent in both market segments.

Retail trade is dominated by small businesses, but firms with at least ten employees account for 50% of all turnover. In addition, there is a tendency for the number of small workplaces to diminish and the number of large ones to increase. What has happened since the mid-1970s is that central department stores have lost market shares and the number of such units has diminished, whereas out-of-town hypermarkets have increased both their market shares and their numbers. This development is expected to continue, assuming that town-centre locations are not made more accessible to motor traffic. Shops in semi-central locations, at all events, are expected to be the main losers. Thus, demand for premises will probably focus on out-of-town locations, in which case it will mainly refer to non-convenience goods. In the convenience goods sector, low-price stores are likely to take market shares from others (Jacobsson 1994).

Wholesale trade is also dominated by small businesses. This is a very heterogeneous sector, but, in terms of turnover, businesses with more than 50 employees account for half the trade. Wholesale trade is also very much a big-city phenomenon and is expected to remain so. Its need for warehousing facilities, however, appears to be diminishing, the tendency being for smaller quantities of goods to be kept in stock (Jacobsson 1994).

5.7 Industrial properties

Industrial properties are a heterogeneous group. They can take the form of buildings and facilities suitable for a wide spectrum of activities, but they can also be suited to one specific activity alone. Advance factories are one instance of the former category, pulp mills of the latter. Midway between the two, of course, there are properties that can be applied to a new purpose after more or less extensive alteration, but an estimated 90% of all developed industrial properties are sufficiently flexible to accommodate new activities without very extensive alterations (LMV 1993d).

Two thirds of all industrial buildings have been erected since 1960, but only half of the volume of buildings has been produced in this period. Eighty-five per cent of the buildings are heated, whereas the rest are cold warehousing (Wilhelmsen & Lindgren 1993).

In 1992 there were over 75000 industrial units (Table 5.5), of which 12000 were undeveloped plots (these figures refer to tax assessment units but probably give a fair indication of the property stock). Of these units about 26000 were used for manufacturing, including about 7000 for metal and machine manufacturing and 5000 for woodworking industry. Not quite 14000 were used as warehousing and about 10000 as petrol stations and repair shops. In addition there were 6000 "industrial units with other buildings", which can mean, for example, an office building on an industrial property.

Most industrial properties are owned by natural persons and companies. Natural persons own mainly small units, and so the value of their holdings is limited. Companies are the largest owner category in value terms. In 1988, for example, private individuals owned one third of all units but only 5% of the property

Table 5.5 Number of industrial units (tax assessment units), 1992. *Source:* SCB (1993i).

Category	No.
Plots	12000
Chemical industry	700
Food industry	1600
Metals, machinery	7000
Textiles, clothing	800
Wood-working	4900
Other factories	10600
Advance factories	1200
Industrial units with other building	5600
Warehouses	13700
Depots, parking spaces	900
Industrial units with renewal building	500
Petrol stations	3000
Repair shops	6800
Power production units	1400
Quarries, etc.	5100
Total	75800

value, whereas companies owned about 40% of the units but 60% of the property value. The Swedish government was a large capital owner, controlling 25% of the entire stock of industrial properties, although the number of units owned was limited (Mattsson 1992). A large part of government capital is invested in power stations.

Turnover of industrial properties on the open market in 1992 was about 1000 properties, 600 of which were intended for manufacturing and 200 each for petrol stations and warehousing. In addition, some 1300 acquisitions were made through internal purchase, i.e. with community of interest between buyer and seller (SCB 1993e). Over and above direct property purchases, corporate acquisitions can take place through the transfer of shares in a company that owns a property (LMV 1993d). The volume of these transactions is not known. Property turnover in 1992 was almost at the same level as in 1985, the time when industrial property prices began rising. At its peak in 1988, turnover was twice the figure for 1992 (SCB 1986–93, *Reports Serie P*). One tentative, partly hypothetical conclusion to be drawn from this brief period is that higher prices and increased turnover went together. This also tallies with previous accounts of the house and rental housing markets.

The movement of industrial property prices, it is true, resembles that of office prices, but it started a few years earlier when industrial investment began to gather speed. Prices rose in both nominal and real terms until 1990, after which they fell. These changes, then, primarily reflect not the general movement of property prices but, if anything, the changes occurring in industrial investment (Mattsson 1992). The decline in industrial output has since contributed towards a surplus of premises, which has reduced both property prices and rents.

The Central Office of the National Land Survey (LMV 1993d) has carried out a detailed study of price movements from the mid-1990s to the New Year 1993. This suggests that the fall in prices was somewhat greater than shown by the official statistics, namely about 30%. Prices fell most in Stockholm and Göteborg (about 50%) and least in the sparsely populated northern areas (about 10%). Elsewhere in Sweden they fell by about 30%. Thus, it was the areas with the largest increases in the 1980s that experienced the greatest falls. The fall in prices does not appear to have varied appreciably with the type of industry, which is perhaps not very surprising, considering that most of all industrial buildings can be converted to other activities.

Industrial property price levels, of course, vary a great deal, depending on the part of the country in which an industrial property is located. Table 5.6 conveys an approximate picture of the average values of newly completed premises of normal standard, developed land included. The highest levels, not unexpectedly, occur in Stockholm and Göteborg. Value then declines with the size of locality, and the lowest values are found in the sparsely populated rural areas. In these areas the value of the buildings does not normally account for more than 10–30% of as-new value, and so more often than not it is uneconomical to put up a new building instead of buying or renting existing premises, unless building development qualifies for State grants or other forms of aid.

Another fact observable from Table 5.6 is that hardly any premises at the moment attain a value corresponding to their replacement cost. This can be put at SEK6000–9000 per m^2 for basic industrial premises, not including land and infrastructure connections. The lowest price levels used to be confined to the interior of Norrland, but they are now also observable in sparsely populated rural areas of southern Sweden, partly because the prefabricated housing and woodworking industry of southern Sweden has been badly hit by the fall in housing production (LMV 1993d).

Table 5.6 Market value level of newly completed manufacturing facilities, 1992 (SEK per m^2 usable floor area). Source: LMV (1993d).

	SEK per m^2
Stockholm	5000–6000
Göteborg	4000–5000
Larger towns/cities	3000–4000
Other country towns	2000–3000
Medium conurbations, central Sweden	1200–2000
Countryside, most of Sweden	700–1200

Price differences are also connected with the age and type of premises. A building from the 1960s or earlier is judged to have a market value 30–40% lower than that of a newly erected building. A newly completed warehouse is judged to have a market value equalling 60–80% of that of a newly completed manufacturing facility of comparable quality.

The emergence of advance factories (1200 in number, according to Table 5.5) is an interesting phenomenon. These are properties that companies can rent for their operations. Small and newly established businesses in particular may need this arrangement, to spare them having to tie up capital in property investments. Advance factories can be anything from small, plain buildings to large, exclusive complexes. Often they are designed for both industrial and office use. Compared with other industrial buildings they are relatively valuable. Roughly a quarter are located in the Stockholm region, and this quarter represents nearly half the total value of all advance factories in Sweden (Mattsson 1992).

People in the business are convinced that the industrial rental market is on the increase. During preparations for the 1988 property tax assessment, it was found that a quarter of all manufacturing facilities, industrial offices and warehousing were partly or wholly rented (Sundqvist 1987). This means that the rental market for industrial premises is much larger than the number of advance factories indicates and that the latter are just one among several kinds of rented premises.

The average rent levels for industrial premises describe much the same geographical pattern as building values (Table 5.6). The spread of rents within a locality, however, can be considerable, depending on the standard of the premises. The average rent at the beginning of 1993 was about SEK500–700 per m^2 a year in the Stockholm region, whereas in other large and medium localities it was normally about SEK300–500 per m^2. The direct yield on the property value usually varied between 10% and 14%, which is appreciably higher than a few years earlier. The yield was lowest in the Stockholm region (10–12%). Rents in Sweden are at the same nominal level as three years ago or even slightly lower. This is because of a glut of premises, usually amounting to 5–10% in the large and medium localities (Fastighetsvärlden 1993, SML 1990, 1993). SEK50–100 per m^2 is estimated to go on running costs, and the remainder on the servicing of capital.

5.8 Tax-exempt units

Some properties are exempt from taxation, which means that they are not tax-valued either. Tax exemption is based on the property being deemed to confer such a benefit to the community as to merit tax exemption. Urban properties in this category mainly comprise buildings used for certain special purposes and buildings owned by certain institutions. Land belonging to the buildings is also tax exempt. It is perhaps doubtful whether this type of property should be included in a chapter dealing with the property market. If, however, the current privatization tendency becomes widespread, parts of the stock may come on the market. The account given also provides an indication of what is considered to be activity beneficial to the community.

The categories and numbers of special buildings are shown in Table 5.7. Two of the building categories may require closer explanation. Distribution buildings are buildings included in gas, heat, electricity or water distribution networks. A public building is a building belonging to public authorities or a community of some kind and used for general administration, the administration of justice, public order or safety. This group also includes leisure centres.

The table speaks for itself and hardly requires any further comment. For the

Table 5.7 Tax-free special buildings (tax assessment units), 1992. *Source:* SCB (1993i).

Category of building	No.
Distribution	12300
Wastewater processing plants	6500
Central heating plants	300
Car service	10100
Bathing, sport and athletics	8400
Schools	9900
Cultural	6000
Ecclesiastical	14700
General	6100
Communications	10800
Defence	1800
Other public	1000

record, however, it should be mentioned that, in addition to the special buildings included in the table, there are also other tax-exempt buildings belonging to certain institutions specified by law.

Installations such as roads, railways, streets, power lines, gas mains, and telephone lines are completely excluded from tax assessment, are included in parent properties or are entered at symbolic values only, in spite of representing great capital values. The value of these installations has been estimated at SEK500000 million.

CHAPTER 6
Case studies

In this chapter we refer back to the models of the development process that were presented in Chapter 4.3. Thus, the cases illustrate developments on both privately and municipally owned land, involving both professional and non-professional developers. A case study is also presented concerning implementation of the joint land development procedure. The case studies in the final section describe the procedure for the purchase and sale of properties and tenant-owner titles. The property sales involve single-family dwellings, apartment buildings and a commercial property.

6.1 Barnsjö, municipality of Tyresö

The municipality of Tyresö
Tyresö is a typical suburban municipality about 15km southeast of the centre of Stockholm, most of whose residents commute to Stockholm or other localities.

The municipality has expanded very substantially in terms of both population and building development over the past 30 years. In 1960 it had just over 5000 inhabitants, whereas today it has more than 35000. This expansion has been based to a large extent on the renewal of older single-family and secondary housing areas. Nearly half the municipal population are house-dwellers.

Barnsjö
The development area comprises about 30 hectares. An older type of detailed plan (*byggnadsplan*) dating from the beginning of the 1950s reserved this area for development of second homes. The maximum permissible building area was $70\,m^2$ and the minimum plot size $2500\,m^2$. At the end of the 1980s the area had 18 plots, most of them with second homes of a relatively low standard. The plots were privately owned, except for four that had been acquired by the municipality during the 1970s. The municipality also owned the green area surrounding the properties (see Fig. 6.1).

New planning
The comprehensive plan adopted by the municipality in 1985 designated the development area as a recreation area with pre-existing second homes. There was, however, a pent-up need for building development in the area and at the end of the 1980s the property owners expressed the desire for a new planning

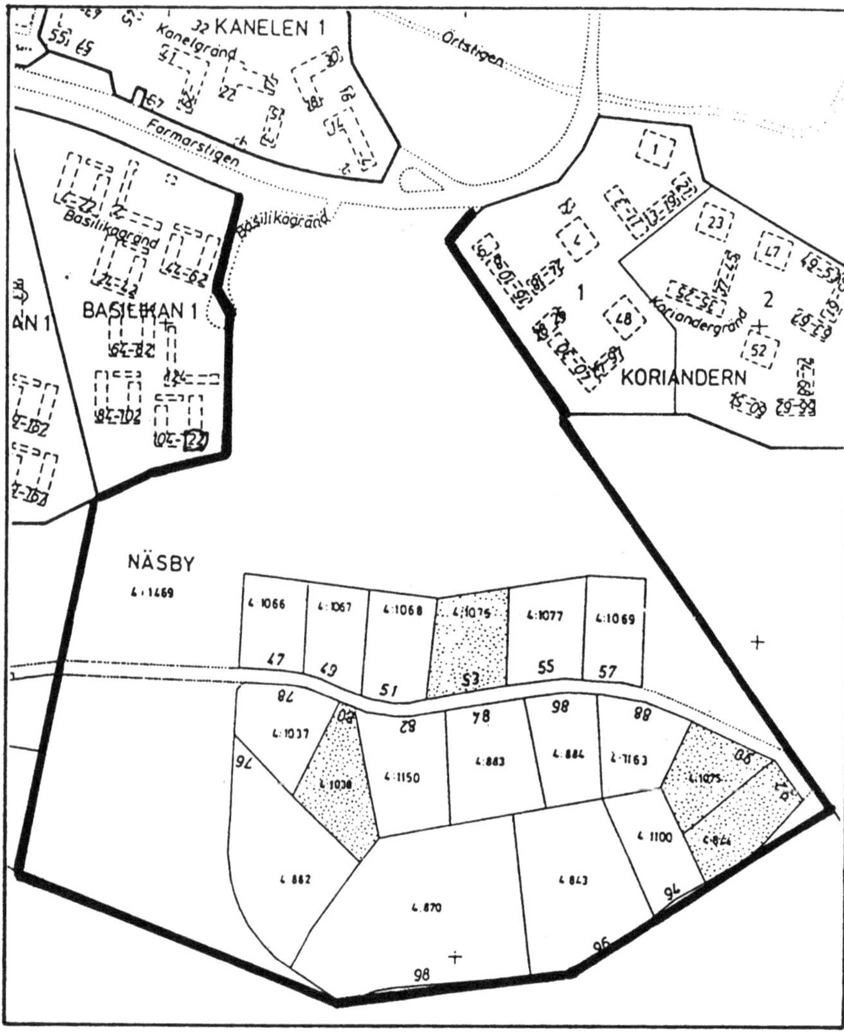

Figure 6.1 The development area at the end of the 1980s. Municipal land holdings are shaded.

policy to allow the development of the area for permanent residence with modern streets and water and sewerage mains.

The municipality therefore began work on a new detailed plan in the spring of 1989. A draft plan was presented in July 1989 and during the late summer of that year consultations were held with the property owners. After minor revisions, the detailed plan was adopted by the municipal council in February 1990, acquiring the force of law in March.

The new detailed plan (Fig. 6.2) contains three types of areas:

- An area of about 60 dwelling units in apartment buildings. These buildings might comprise, at most, two storeys or one storey and a converted attic. The plan limited the total building area to a maximum of 4200 m^2. Furthermore, the buildings were to be of small-scale design.

- An area of detached or semi-detached housing development. To regulate building density, the plan stipulated a maximum number of properties for each block and a maximum building area (160 m^2 for the main building and 40 m^2 for a shed, garage or suchlike ancillary buildings). Altogether this allowed an infill of from 12 to about 30 properties.
- Areas for public spaces including streets and natural open ground.

The implementation period for the detailed plan was set at 15 years.

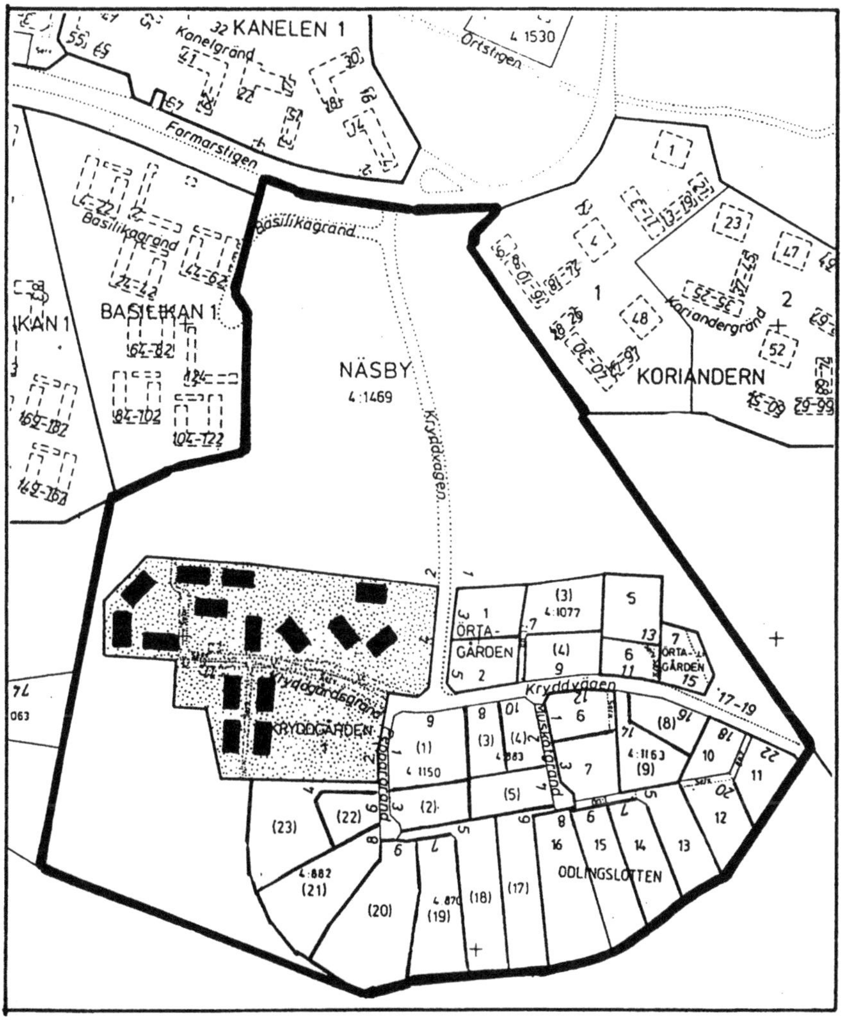

Figure 6.2 Land use in the detailed plan. Multi-family housing areas are shaded.

Responsibility for implementing the plan

Multi-family housing The apartment buildings were to be constructed by one developer. This assumed transfer to single ownership of land within the block, most of which belonged to five private landowners and the municipality. This was accomplished in the following way. Parallel to the preparation of the detailed plan, negotiations began between the plot owners and the municipality. The negotiations led to a swap agreement whereby the municipality obtained the land for multi-family housing development and the property owners, in exchange, acquired plots owned by the municipality in the single-family housing area. Once these land transactions had been completed, the municipality was able to adopt the legally binding detailed plan.

Next, after the detailed plan had been adopted in August 1990, the municipality entered into a land allocation agreement (*markanvisningsavtal*) with the developer, HSB-Stockholm. Under that agreement the municipality was to convey the land to HSB at a price of rather more than SEK1 million. The agreement also laid down that HSB was to pay charges totalling SEK3 million to the municipality for streets and water and sewerage mains (see below). The new homes were to be financed with State housing loans and the municipality reserved the right of referring buyers to 50% of the dwellings. Otherwise, this agreement supplemented the provisions of the detailed plan. Among other things, it provides that HSB was to have about 60 (not more than 66) tenant-owner flats built, mainly in accordance with a more detailed plan appended to the agreement. The more exact design of the building development was to be decided in close co-operation with the municipality, the aim being for only natural, well proven materials to be used for the building. The buildings were to be constructed in such a way as to facilitate pre-separation of refuse and waste, and the buildings were to be connected to the municipal district heating network. The agreement also included timetabling provisions. Building development was to proceed in accordance with a timetable approved by the municipality that provided for the area is to be fully developed not later than August 1992, in spite of the implementation period fixed in the detailed plan of 15 years (1990–2005).

A building permit was applied for and granted in February 1992, and the development – ultimately comprising 60 tenant-owner flats – carried out by HSB in the spring of 1992. Partly because of the recession, however, not all the flats have been sold. HSB still has about one third of them on its hands.

Single-family housing To settle the future division of the block into new property units, the municipality drew up a subdivision plan (*fastighetsplan*) parallel to the detailed plan. As was explained in Chapter 3.1, this plan then forms the basis of the new property subdivision. In this particular case the subdivision plan meant that land would be parcelled off from the existing property units and that, in certain cases, land would be transferred from one property unit to another.

Responsibility for initiating land subdivision and carrying out the building development devolves on the individual property owners. Today about one third of the development sites have been built on. As stated earlier, however, the implementation period for the detailed plan does not end until 2005, and, given a general economic improvement, the plan is likely to be fully implemented.

Streets and water and sewerage mains Responsibility for the construction of these communal facilities devolves on the municipality. As was shown in Chapter 3.1, the municipality is then entitled to obtain coverage for its costs by levying charges on the property owners.

Where roads were concerned, an inquiry was conducted into ways in which these charges could be levied parallel to the compilation of the detailed plan. The total road costs, including remuneration for the land needed for them, came to just over SEK3.2 million. Of this, the distributor road in the northern part of the area cost SEK1.5 million and the remaining streets directly linked with the building development cost SEK1.7 million. These were divided between 60 dwelling units in multi-family buildings and 30 units in the form of detached and semi-detached houses. The distributor road was then expected to serve all dwellings to an equal extent. In other words, the multi-family and single-family dwellings respectively were to contribute two thirds and one third of SEK1.5 million. The remaining SEK1.7 million, i.e. the cost of the other roads, was basically apportioned with reference to the plot area of the two types of building development (22000 m^2 and 42000 m^2, respectively). This meant the multi-family and single-family buildings paying, respectively, one third and two thirds of this cost. Finally, as regards the apportionment of costs between the 30 house owners, it was decided that costs should be apportioned as equally as possible, having regard to the extent to which the existing property units had paid charges for the older streets. It was therefore proposed that new property units or new parcelled properties in the plan should pay 50% more than the original property units. Summing up, the inquiry concerning street cost charges resulted in the multi-family housing area having to pay about SEK1.6 million (or on average SEK27000 per dwelling unit) and an original property and a new single-family property paying, respectively, about SEK42000 and 63000.

Charges for water and sewerage facilities are based on a tariff applying to the municipality as a whole. The Tyresö municipal tariff is based on three parameters:

- A charge depending on the number of points at which the property is connected to the municipal water and sewerage network. Normally there is only one such connection.
- A charge based on the number of dwelling units in the property.
- A charge related to the area of the property. For the multi-family housing area, this meant a charge of SEK1.3 million. Where the single-family properties were concerned, a property of 900 m^2, for example, would in 1991 have paid a charge of SEK84000. For a property measuring 2000 m^2 the charge was SEK105000.

Concluding reflections

The Barnsjö area is among other things an example of the municipality originally taking a passive but not negative line on development. The municipality realized that development of the area would probably take place sooner or later, but it did not take any active measures to initiate development. The comprehensive plan therefore indicated that the area was to be preserved in its original state, with second homes. Not until the property owners in the area took the initiative to bring about renewal and infill of the area did the municipality take action and embark on new detailed planning, partly because this opened up the possibility of bringing about socially desirable multi-family housing development.

The Barnsjö development also points to the importance of being able to change property and ownership structures so that a project can be completed in a suitable manner. In the multi-family housing area, the municipality took upon itself the responsibility of acquiring the land (before transferring it to a developer). Changes of property subdivision in the single-family housing area were based on a property division plan that makes it possible for land to be compulsorily transferred between property units if the property owners are unable to agree among themselves.

Finally, it is demonstrated that in principle all development costs had to be borne by the property owners. Although the municipality was primarily responsible for the construction of streets and water and sewerage mains, it was able afterwards to charge the property owners for the cost of these facilities. The only subsidies accruing to the property owners were the interest contributions linked to the State housing loans (see Ch. 3.2).

6.2 Södra Sticklinge, municipality of Lidingö

The municipality of Lidingö

The municipality of Lidingö is an island bordering on its southwest with the municipality of Stockholm. Lidingö is generally looked on as an "exclusive" residential location, partly because of its attractive scenery and its proximity to the central parts of the Stockholm region (about 5–6km from central Stockholm). The municipality has nearly 40000 inhabitants and over 60% of its gainfully employed population of about 22000 commute to other municipalities.

During the 1940s, Lidingö was a typical garden city. Several large multi-family housing developments have been added since then, and at present there are something like 18000 dwellings, of which about 60% are apartments and about 40% single-family dwellings.

Södra Sticklinge

The development area totals about 250000m^2. Most of it occupies a rocky plateau, which today is dominated by a combination of rock, spruce and pine. This was a greenfield site. To the north of it is the Norra Sticklinge development dating from the 1970s and 1980s.

An area development plan or detailed comprehensive plan (*områdesplan*) from 1982 provided for multi-family and single-family housing development in the area in order to supplement Norra Sticklinge and to provide a population base for such things as schools and commercial services.

In April 1986, the JM Byggnads och Fastighets AB construction and property company purchased from the municipality about 65000 m^2 within the development area. This acquisition formed part of a swap between the developer and the municipality in which, among other things, the developer transferred another potential development area to the municipality. At the same time a type of prior development agreement (*optionsavtal*) was concluded between developer and municipality, in which the municipality pledged itself to transfer later on, in January 1988, another 70000 m^2 or so of land at a price of SEK4 million. When this purchase was completed at the beginning of 1988, two important conditions were added to the contract of sale: the developer was to carry out the planned housing development in accordance with a programme for an architects' competition adopted by the municipality in November 1987, and the developer was to engage the architect selected as a result of the competition.

In the competition, the municipality and developer invited four firms of architects to submit schemes for the development of the area. Basically the competition brief was to show how a development comprising about 300 dwelling units, 250 apartments and 50 single-family dwellings could blend with the pre-existing environment. The competition was decided in June 1988, and the winning entry came from the FFNS Arkitekter practice under the watchword "gently" (*varsamt*).

The scheme was processed and in January 1989 the municipality declared that it was to form the basis for subsequent work on the detailed plan. In December 1990, after further processing of the scheme, the municipality approved a programme for the detailed plan in which it was said that the construction of upwards of 360 dwelling units would form the basis of a detailed plan and contract negotiations between the municipality and the developer. In other words, one effect of these processing rounds was to increase the intensity of development by about 20%.

The detailed plan

The practical work involved by the detailed plan was done by FFNS Arkitekter in consultation with the municipality, whereas the municipality was responsible for drawing up the detailed planning map and, of course, for the formal processing of the planning transaction (consultations, exhibition and approval of the plan). The final draft version of the plan provided for the erection of 24 four-storey apartment buildings, with a total of some 270 apartments, and 101 two-storey, single-family dwellings (breaking down into 44 terraced houses, 19 semi-detached houses and 19 detached houses). All in all, the plan implied the construction of some 370 dwellings. As regards tenure, the multi-family homes were to be partly rented and partly tenant-owned, whereas the single-family dwellings were to be freehold.

In the northern part of the development area there were good opportunities for linking up with pre-existing roads; the new development is served by a large traversing street having two connections with the pre-existing street network. The same goes for water and sewerage mains and electricity. It was assumed that heat would be supplied from an oil-fired central unit within the area. Connection to the municipal district heating network was judged prohibitively expensive because of the length of pipe-laying this would involve. The plan also provided for two day nurseries and for some leisure amenities incorporated in certain of the apartment buildings.

The implementation period for the plan was set at 15 years. The detailed plan was adopted by the municipality in December 1991 and acquired the force of law in January 1992.

The detailed plan is very detailed concerning, for example, the location of buildings. As mentioned in Chapter 3.1, building permits and property subdivisions must be in accordance with the detailed plan, except for "minor deviations". When the implementation of the detailed plan began to be studied more closely, however, it became apparent that the plan was not entirely practical as regards the proposed subdivision of properties, the positioning of certain buildings and the boundary between developed sites and public places. At some points in four precincts, boundary changes had to be made in the plan of the order of 2–7 m. The municipality took the view that these changes were more than "minor deviations", and so a new detailed plan was drawn up and adopted in May 1992. Figure 6.3 shows parts of a precinct in the original and the amended detailed plans. This illustrates the adjustments that had to be made to the positioning of buildings and to the boundary between developed sites and public places.

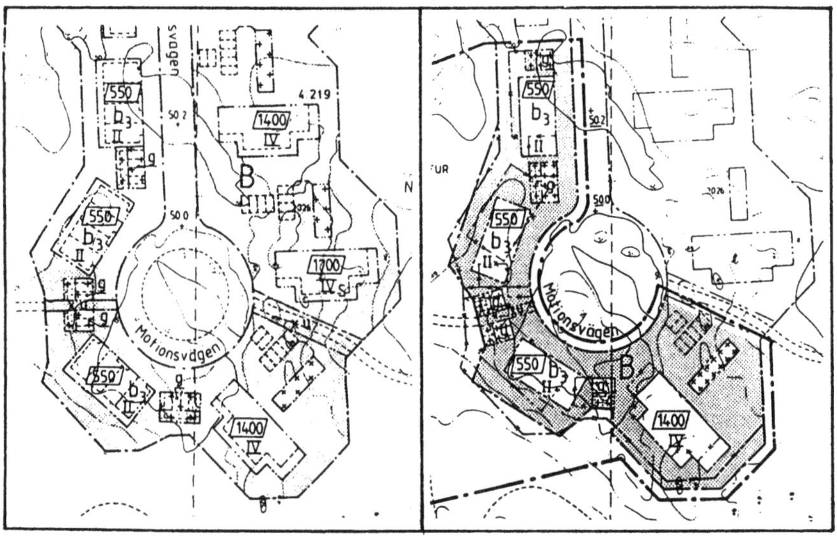

Figure 6.3 Examples of changes in the detailed plan.

Responsibility for implementing the plan

One precondition for the adoption of the detailed plan by the municipality was that a land development agreement (*exploateringsavtal*) had to be concluded between the municipality and the developer. Negotiations for this agreement proceeded parallel to the preparation of the detailed plan, and an agreement had been drafted by June 1991. The agreement was finally signed in December, just before the adoption of the detailed plan.

The main terms of the agreement can be summarized as follows:

- As stated earlier, most of the land to be built on was owned by the developer. Partly as a result of the processing of the development proposals and the increased intensity of development, however, the developer needed to acquire a certain amount of additional land from the municipality. This mainly concerned four precincts in which detached houses were to be built, and the price of this land was SEK7 million.
- The developer was to transfer to the municipality, free of charge, land for streets and parks and development sites for day nurseries, a leisure centre, a central heating plant and transformer substations.
- The developer was to construct and pay for streets and water and sewerage systems within the area.
- The developer was to pay the municipality a contribution towards the share of the development area in the main system for water supply and sewerage outside the area. This contribution amounted to SEK3.2 million.
- The municipality in turn was to reimburse the developer for street, water supply and sewerage installations serving the above-mentioned day nursery and leisure centre areas. This reimbursement totalled SEK1.7 million.
- The developer undertook to construct housing and other installations within a period of not more than four years from the detailed plan acquiring the force of law.
- About 160 of the apartments were to be occupied on a rental basis and about 110 as tenant-owner homes. All single-family dwellings were to be conveyed freehold.
- The building development was to meet the requirements for national housing loans. In addition, a certain proportion of the dwellings was to be allotted through the municipal housing exchange. (This proportion varied according to tenure.)
- Lastly, the validity of the agreement was made conditional on the detailed plan acquiring the force of law not later than March 1992.

Between June 1992 and March 1994 building permission was granted for the six precincts comprising multi-family housing development. Most of this housing has been completed. What remains consists of four precincts of detached housing.

Concluding reflections

The development process in Södra Sticklinge must be termed typical of the boom in the second half of the 1980s. The design of the building development was to some extent regulated through negotiations between municipality and developer. The number of dwellings, for example, was raised from the original 300 or so to, finally, 370. This increase in the intensity of development was primarily in the developer's economic interests, but it also conferred a certain economic benefit on the municipality. For one thing, the municipality was able to charge more for the land transferred to the developer in 1992, and for another thing this increased the contribution received by the municipality from the developer for the development area's share in the main water supply and sewerage systems.

The content of the land development agreement basically complied with the "normal model", i.e. the municipality did not try to impose on the developer any economic or other obligations beyond those that could be achieved with the support of legislation (see Ch. 3.1 above). During the boom period of the 1980s, however, there were several instances of developments in which municipalities harvested some of the profit that ought legitimately to have accrued to the developer. Mostly, however, this applied to developments for commercial purposes. In the case of Södra Sticklinge, it should also be noted that the land development agreement was concluded at the end of 1991, by which time housing construction had entered a period of crisis. In other words, there did not really exist any sizeable profits on the development that the developer was ready to share with the municipality.

6.3 Uddaberg, municipality of Skövde

This case study provides an example of development implementation supported by the Joint Land Development Act (*lagen on exploateringssamverkan,* ESL; see Ch. 3.1 above).

The municipality of Skövde
Skövde, with a population of not quite 50000, is in the south of Sweden, on the railway between Stockholm and Göteborg. It has a relatively diversified economy, with industries (Volvo among them), a college, a county hospital, military units and public administration.

Uddaberg
The development area occupies the eastern slope of a hill, Billingen, designated as being of outstanding scientific and recreational interest. This is a very attractive location for building development, and it directly adjoins areas planned and developed previously.

The existing land use comprised agricultural land and permanent housing on large plots. Something like 80% of the land was undeveloped. The area originally comprised about 20 properties (although this number had been slightly reduced through amalgamations). The greater part of the area was owned by two construction companies: Optimalhus AB and Sjödalshus AB (the shaded proper-

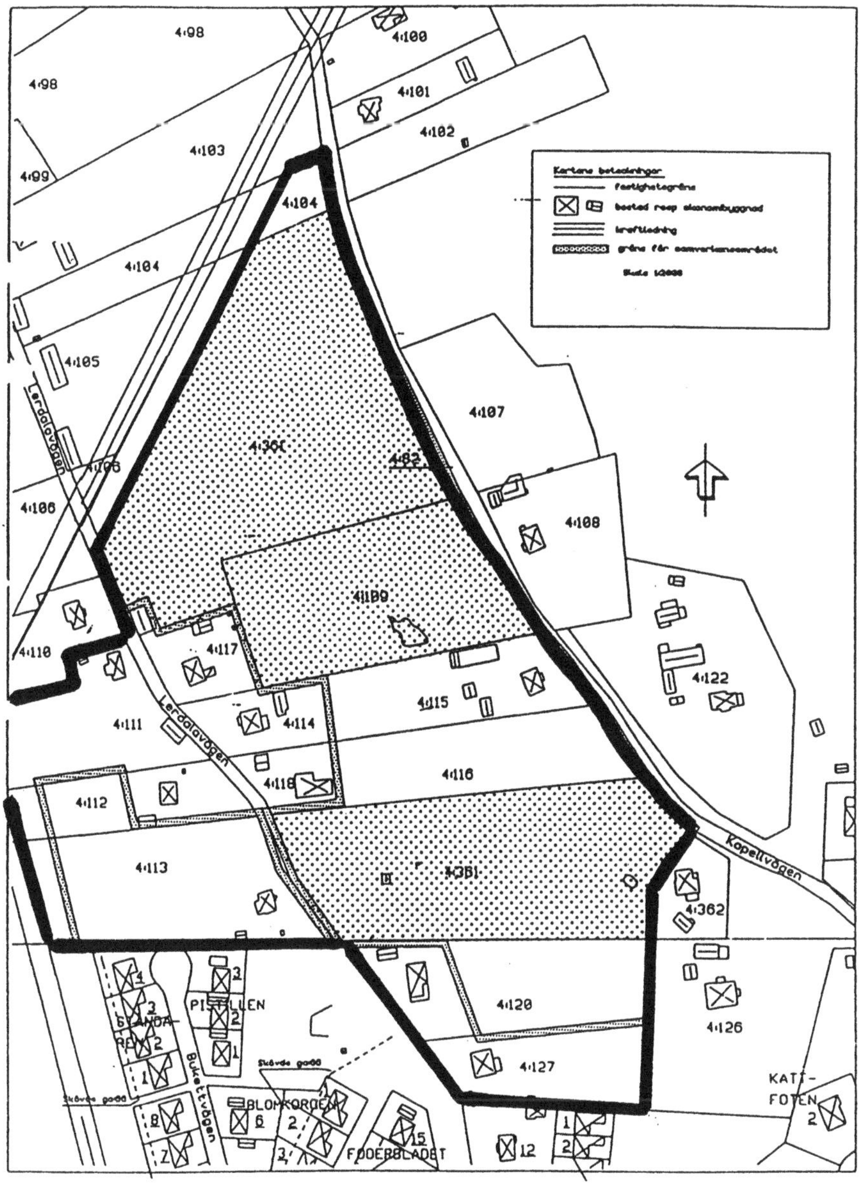

Figure 6.4 The development area, with the developing companies' landholdings shaded.

ties in Fig. 6.4). Otherwise, the properties were privately owned and so the municipality had no proprietary interest in the area.

Why joint land development?

The comprehensive plan stipulated that the area was for single-family housing development only. It also indicated that co-ordinated development under ESL could be appropriate. The development area is in fact a good example of the kind of situation where joint land development is intended. First, the property owners were interested in playing an active part in planning and developing the area. Secondly, the property and ownership structure of the area was fragmented, which meant that the development would require restructuring of the existing property subdivision. Thirdly, the municipality had no intention of taking part in this restructuring by acquiring development sites in the area. Joint land development, in other words, was mainly prompted by the rules of ESL concerning profit sharing and reallotment (see Ch. 3.1). On the other hand, it was not the intention for the property owners to be jointly responsible for the construction of streets and water and sewerage mains before handing them over to the municipality. This development responsibility would devolve on the municipality.

New planning

The project was initiated in connection with one of the construction companies (Optimalhus AB) conveying to the municipality its desire to be allowed to plan the area under its own auspices. This led the municipality to consider applying the joint land development procedure to the area for reasons already stated.

Special area regulations One initial step in joint land development is for the municipality to indicate, through special area regulations (*områdesbestämmelser*), that joint land development may take place in a certain area. The municipality began work on special area regulations in the autumn of 1990. These regulations were adopted in May 1991, acquiring the force of law in June. In the special area regulations, the area where joint land development was to take place (the joint development area or *samverkansområde*) comprised only part of the area for which it was intended to draw up a detailed plan later on. This restriction was prompted mainly by the desire of the property owners to take part in joint land development.

Development decision I The next stage in joint land development is for the Property Subdivision Authority to decide certain questions relating to the development (*exploateringsbeslut*). In June 1994, therefore, the other construction company (Sjödalshus AB) applied for joint land development proceedings. In September 1991 the Property Subdivision Authority made an order whereby the joint development area was reduced in relation to the special area regulations. Developed plots in the eastern part of the area were not included because the property owners did not want to take part and the development could be accomplished in a suitable manner without their participation. Within the joint development area, two property owners chose not to take part in the joint development. They comprised an individual property owner and Optimalhus AB, which had originally initiated the project but was now insolvent. The Property Subdivision Authority made a formal order whereby the other property owners were to set up a joint land development unit (*exploateringssamfällighet*).

The municipality played an active part in the development project during the introductory phase of the development process. Several consultative meetings were held at which the municipality informed the property owners about the preconditions and aims of the intended development, such as the floor-space index, the type of building development, responsibility for communal facilities and the rules of ESL. The attitudes of the property owners to development during the introductory phase can be termed "reservedly favourable". In January 1991 the property owners set up a working party to pursue the project further by carrying out design work and planning of development sites. In this way the property owners in general, and Sjödalshus AB in particular, became more active. Prior to the development decision by the Property Subdivision Authority, the working party had already prepared a sketch plan that the municipality found acceptable as the basis of a definitive detailed plan. The property owners, aided by a consultant, then went ahead with planning work, in consultation with the municipality.

Development decision II At the beginning of March 1992, the municipality approved a draft detailed plan for consultation and display. That plan implied something of a modification of the terms of joint development. The plan stated that there was no reason for certain properties on the fringe of the development area being included in the joint land development area. The Property Subdivision Authority's development decision II, at the end of March, had the effect (with the property owners' consent) of further reducing the size of the joint development area. Four of the properties were now to be included in the joint development unit, whereas two opted out.

The members of the joint development unit were agreed that the profit from the development was to be apportioned in accordance with the main rule of ESL. That is, the profit was to be allocated according to the *area* of the property units. The participatory shares of the four properties were computed as follows:

Property unit	Area (m^2)	Share (%)
4: 115	3390	10.7
4: 116	3950	12.5
4: 120	2950	9.3
4: 361	21360	67.5

It is worth mentioning that the owner of 4: 361 (Sjödalshus AB) acquired property 4: 115 on a voluntary basis after the development decision but before the detailed plan had been adopted. The Property Subdivision Authority's decision also meant that the two properties not included in the joint development unit (4: 104 and 4: 109) were made subject to compulsory purchase by the other property owners.

Detailed plan The detailed plan was adopted by the municipality in September 1992. The detailed planning area conforms to natural boundaries. The area is bounded in the south by a previously planned area, to the west and north by a restricted area for a power line, and in the east by a road. The area east of the road is being planned for other forms of development, namely about 160 dwellings in the form of multi-family housing. The area is traversed by a ravine, an environmental asset to be preserved in the event of development. The purpose of the detailed plan, according to the planning description, was to form the basis for constructing about 60 dwellings, most of them in single-family housing groups. About ten pre-existing detached houses were incorporated in the plan. The joint development area, however, includes only parts of the total planning area (see Fig. 6.5). Building operations within the joint development area itself comprise 36 freestanding single-family dwellings, together with roads, footpaths, lighting, water and sewerage mains and the provision of a green space (the ravine) in the middle of the area.

We may note in passing that the detailed plan was contested by neighbours of

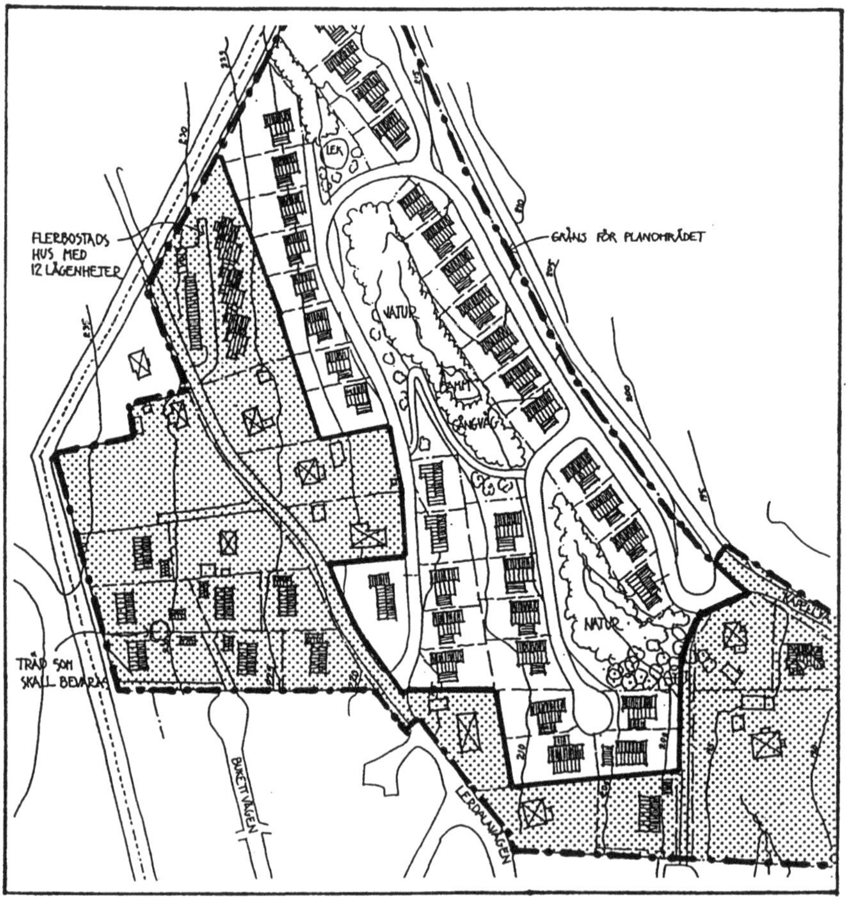

Figure 6.5 Illustration for the detailed plan. The joint land development area constitutes part of the planning area.

the detailed planning area. They objected to the design of a proposed foot- and cycle-path in the plan. The complaints were withdrawn, however, subject to the moped-trap promised by the municipality actually being constructed, which the municipality guaranteed it would. Accordingly, in November 1992, the detailed plan became legally binding.

Responsibility for implementation of the plan

Developed land The Property Subdivision Authority effected an economic settlement and changes of property subdivision in November 1992. As has already been made clear, ESL provides for the profit (appreciation of the properties minus development costs) to be distributed between the properties with reference to participatory shares based on acreage. In other words, the total value of the properties before and after the plan respectively, as well as costs, are to be allotted according to acreage. Before the detailed plan was prepared, the properties in this particular case were judged of identical value per m^2. The main problem, therefore, was to achieve a property subdivision whereby value after the plan would as far as possible be distributed in accordance with the participatory shares. This was achieved by distributing the 36 plots in the plan on the following lines: property 4: 116 received four plots, property 4: 120 three plots and property 4: 361 the remaining 29 plots. This apportionment of value corresponded on the whole to what the properties were to obtain by virtue of their participatory shares (the owner of 4: 361 received about SEK20000 compensation from the owners of 4: 116 and 4: 120). The property subdivision giving rise to this distribution of value is illustrated in Figure 6.6.

Streets, water and sewerage mains In principle, the municipality, referring to development agreements, could have insisted on the joint development unit constructing streets and water and sewerage systems. In the event, however, the municipality opted for constructing these facilities itself. Even so, an agreement was drawn up between the municipality and each individual property owner, showing the street zone to be transferred to the municipality and the charges payable to the municipality by the property owners for road construction. The municipality received the land free of charge, and, for every new plot received under the new detailed plan, the property owners were to pay SEK86000. This meant that properties 4: 116, 4: 120 and 4: 361 were to pay a total of SEK344000, SEK258000 and SEK2494000 respectively, charges being payable when the building permit was granted or when the plot was sold. In addition, the municipality was to receive charges of about SEK60000 per plot for the water and sewerage network.

Concluding reflections

The rules of ESL make it possible for peremptory decisions to be made concerning, for example, land transfer and economic settlement. The Property Subdivision Authority, however, did not need to resort to these coercive rules in

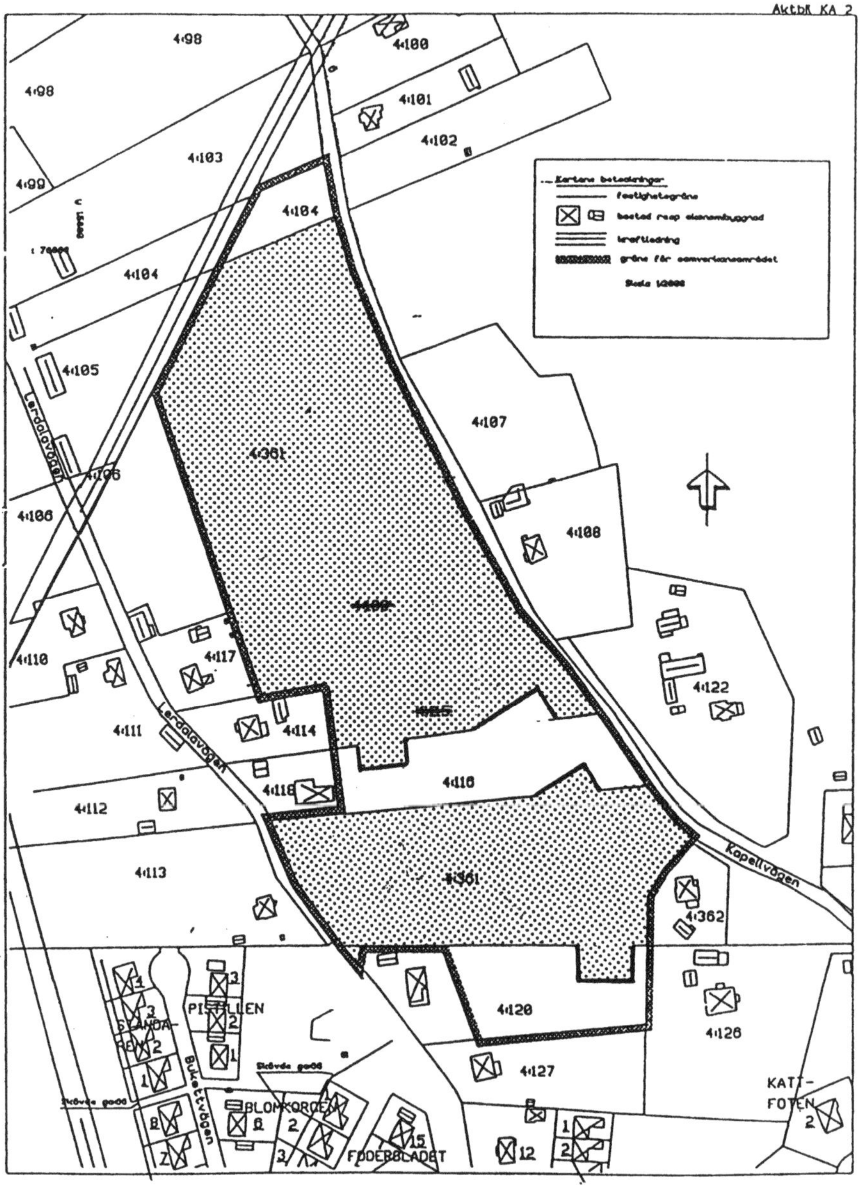

Figure 6.6 Property subdivision in accordance with the detailed plan, prior to land parcelling and the transfer of land to streets (land belonging to the construction company, Sjödalshus AB, is shaded).

practice, because agreements between the property owners could be made the basis for decisions concerning land transfer, payment for the land and the apportionment of costs. Similarly, agreements existed concerning the apportionment of costs relating to public spaces since the co-owners had signed development agreements with the municipality. Only one disputed issue remained. Property 4: 109 had to be purchased (by property 4: 361), but the owner maintained, in the first instance, that the prerequisites of compulsory purchase had not been met, and secondly that the payment awarded by the Property Subdivision Authority was insufficient. An appeal was made against the order but subsequently withdrawn.

Finally, a few words about the present-day situation in the development area. Streets and water and sewerage mains have been constructed by the municipality. As a result of the recession, however, only one plot has been developed. In other words, the area constitutes a reserve of nearly 40 plots ready for building. With the economy taking a slight upward turn, the municipality now sees indications of a resurgence of demand for building plots in the area.

6.4 SAS headquarters municipality of Solna

The headquarters of Scandinavian Airline Systems (SAS) occupy a smart building of glass and plaster stucco from the mid-1980s at Frösundavik, in the municipality of Solna, near a beautiful lake, Brunnsviken, and next to the approach to Stockholm from Arlanda airport. The immediate surroundings are of outstanding quality, with mature broadleaf trees, a green shoreline, old military barracks and other historic buildings. The neighbourhood also includes Haga Palace, which has a large "English park" open to the general public and the ruins of an unfinished palace project, the Royal Palace of Ulriksdal, and several national government offices. Frösundavik also forms part of a more extensive stretch of water and greenery extending far into the central city (see Fig. 6.7).

SAS had already engaged a firm of consultants in 1982 to analyze a handful of possible locations for its headquarters, and the choice had fallen on Frösundavik. There were several factors behind this. The location was strategically appropriate, attractive land was available at the right moment, and the municipal authorities displayed a willingness and ability to overcome the difficulties almost invariably accompanying a major development project.

Most of the SAS development was completed between 1984 and 1988. In terms of expenditure and scope this was a large project, and it was a sensitive one in view of the conservation and historic interests involved. Otherwise, the gestation of the project serves to illustrate development with a municipality as the purveyor of land. The State was also involved, as landowner at the commencement of the project. This is also an interesting case because it was the company's project that governed municipal planning, instead of the other way around. It is true that, when the land was developed, an earlier Building Act was in force, but there is nothing to suggest that things would have taken a different course under the current planning and building legislation.

The following description of the development is based primarily on information from official documents, supplemented by data from articles and interviews (Hultin 1985, Lif 1986, Renlund 1987, 1988). The documents are as follows:

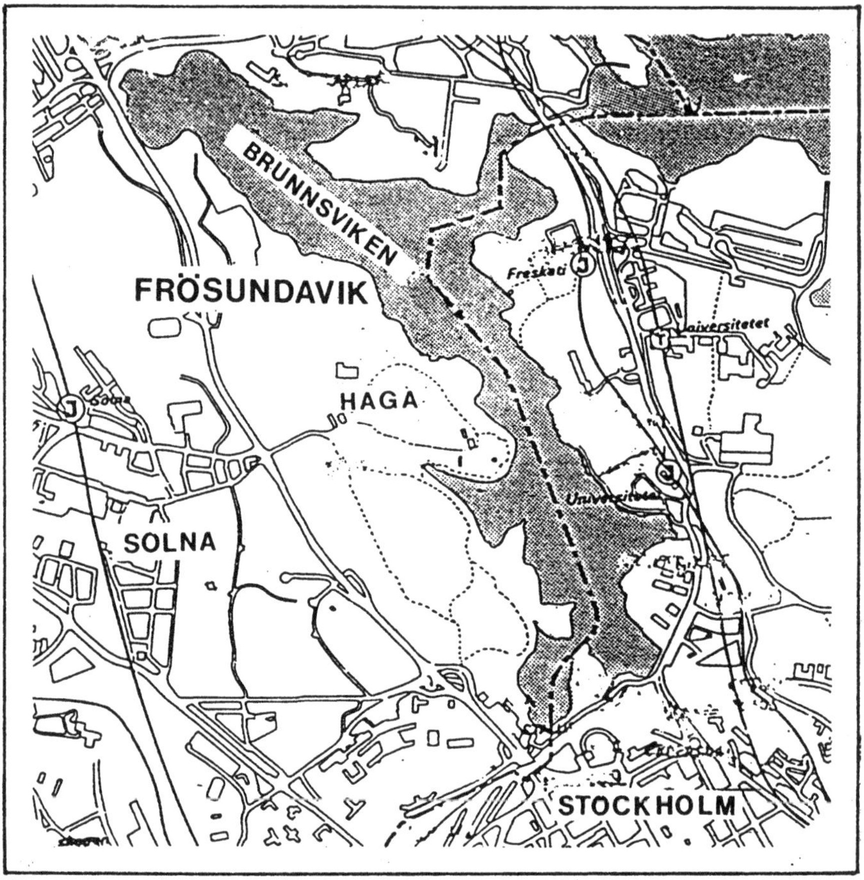

Figure 6.7 Frösundavik and Brunnsviken with surroundings (Source: Solna & Stockholm 1992).

- comprehensive plans, etc. (adopted 1974, 1977 and 1982)
- agreement of principle between the municipality of Solna and SAS (signed May 1984)
- agreement, on the conveyance of real property, and so on, between the State and the municipality of Solna (signed November 1984)
- architects' competition (1984/5)
- area development plan (drafted in February and adopted in August 1985)
- detailed plan (adopted August 1985) and revision of the same (adopted October 1986)
- agreements concerning purchase and development (signed August 1985)
- building permit documents for annexe buildings, etc. (1992)
- comprehensive planning programme (1992–3)

The extent of SAS's land and of the current detailed plan can be seen from Figure 6.8.

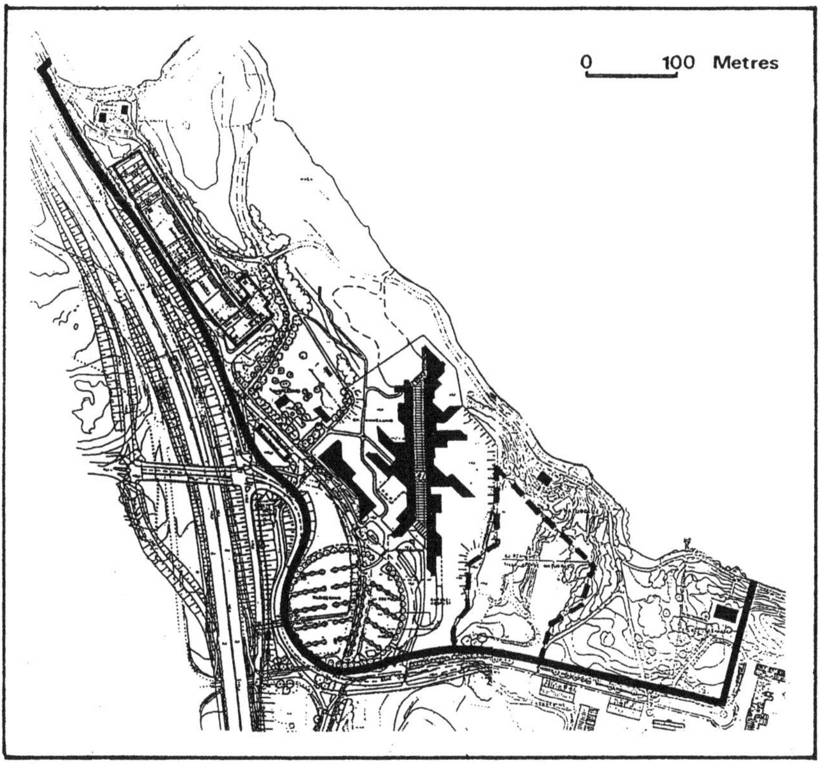

Figure 6.8 The SAS head office at Frösundavik, 1993. Continuous sign: SAS property; dashed line: boundary of SAS property without detailed plan.

Comprehensive plans (1974, 1977 and 1982)

In the 1974 National Physical Planning Programme, the area surrounding Frösundavik had been designated of national interest for heritage conservation, because it contained buildings of historic interest and was of outstanding natural beauty. In its adoption of a comprehensive plan in 1977, the municipality of Solna affirmed that the area formed part of a regional expanse of lake and greenery. In the 1978 Regional Plan, confirmed by the government in 1982, it was stated that Frösundavik was an existing enterprise area that in future would be included in an area for outdoor activity and nature conservation.

Thus, the area had been treated in several comprehensive planning documents, all of which pointed to its importance for heritage and nature conservation. On the other hand, no detailed plan existed, because the original owner, the State, did not prepare detailed plans for its own land before the introduction of the 1987 Planning and Building Act. Development, then, represented a clear departure from the guidelines of the comprehensive planning documents. However, the County Architect, representing the State as supervisor of the planning process, observed during consultations in the spring of 1984 that the establishment of SAS headquarters in the Stockholm region was a matter of such great national and international importance that a departure from the comprehensive plans could be allowed if the buildings were acceptably adapted to the sensitive landscape and if the general public were guaranteed access to the shores of Brunnsviken.

Agreement of principle (May 1984)

An agreement of principle between the municipality of Solna and SAS, in May 1984, defined the conditions on which the land concerned could be sold to SAS. The municipality did not own the land at the time, but negotiations for buying it from the State were well in hand.

The agreement of principle laid down that SAS was to announce a Scandinavian architects' competition for the development. One important requirement would be to show how the office complex could be designed so as to preserve the values of the man-made landscape and heritage conservation and guarantee the general public opportunities for outdoor activity. The development was to proceed by stages, with about 60000 m^2 floor space in the first round and roughly the same amount altogether at later stages. The competition programme was to be drawn up in consultation with the county administration and the municipality.

To avoid anticipating the outcome of the competition, the agreement of principle merely indicated the approximate extent of the land to be purchased. Furthermore, the parties agreed that the municipality was to expand water supply, sewerage and power supply for the area and that SAS was to reimburse the municipality for this at going rates. The municipality undertook, in consultation with the National Road Administration, to plan, complete and pay for the access road from the E4 motorway, as well as roads required during the construction phase. In addition, it was to make vigorous efforts to ensure that the area was provided with public communications.

The agreement also laid down a timetable for subsequent work. The architects' competition was to take place in the autumn and winter of 1984–5. In this connection the municipality was to adopt a detailed plan dovetailing with the winning entry. The writing of a purchasing agreement and development agreement was to be co-ordinated with the adoption of the detailed plan. SAS was to take possession of the land in the spring of 1985 and building was to begin not more than three years afterwards. The construction of technical facilities and road connections was to be synchronized with the planning of building operations. Finally, the agreement presupposed conveyance of the land from the State to the municipality.

Land conveyance (November 1984)

The negotiations for the purchase of land were complicated and involved two transactions. First there was the municipal acquisition of 106 hectares from the State for SEK95 million in November 1984, and then there was the further conveyance of 27 hectares of the municipal acquisition to SAS for SEK45 million in

August 1985 (further to which, see below). It is worth noting that the municipal agreement of principle with SAS concerning conveyance of the land was already signed in May 1984, five months before the municipality acquired the land from the State. The resale to SAS was probably an economic precondition for the municipal acquisition of the land from the State. At the same time, the construction of the SAS office had the effect of making neighbouring land with existing buildings, which the municipality also purchased, more attractive.

The architects' competition (1984–5)

To meet the municipality's demands, and also to find an architecturally interesting and feasible solution to SAS's office requirements, the Scandinavian architects' competition was held in the autumn and winter of 1984–5. Of the 131 groups of architects declaring their interest in taking part, ten practices in Sweden, Norway and Denmark were invited to submit entries in an anonymous competitive round. The published competition programme stated that the building must facilitate the rational organization of office activities, was to be adaptable, must have a small gross area and low running costs, and must be closely integrated with the natural beauty of its location. In addition, SAS took the view that "the world's best airline" must have "the world's best administration building".

The Norwegian Niels Torp A/S practice won decisively with an entry entitled "Prosperity". This entry made maximum use of the terrain, showed great flexibility and had substantial economic qualities. The main building, comprising 55000 m^2, was positioned in the most spoiled part of the area, a former gravel pit. The building was divided into five variously shaped blocks and courtyard units, opening on to the landscape. The units, with their entrances, looked on to a glass-roofed, 180 m-long indoor street of urban character. At the main entrance to the street the blocks were six or seven storeys high, and overlooking the water they were three or four storeys. The indoor street was attractively stepped down towards a small pond, and alongside it more public activities were gathered, such as committee rooms, exhibitions, recreation rooms, dining rooms, and so on. In addition to the main building, a separate low-rise annexe was proposed for SAS group offices, as well as a few other buildings, totalling 15000 m^2, a little to the north of the main complex. A circular parking lot for 400 cars was sited between the motorway and the main building. With minor alterations, this winning entry formed the basis of a detailed plan, a development agreement, building permit documents and contracts.

Area development plan (drafted February and adopted August 1985)

In February 1985 the municipality drafted an area development plan for Frösundavik (an informal but very commonly used planning document in the form of a detailed comprehensive plan), and this was adopted by the municipality in August 1985. It is evident from the planning document that the municipality was still having difficulty in deciding between development or preservation of environmental values. The winning entry, however, was felt to have shown that a large facility could be integrated with the landscape.

The area development plan was intended to provide a basis for the working out of detailed plans and agreements, and also to provide a comprehensive view of the relation of the SAS area to its surroundings. For this reason the area development plan also included the old barracks. Within the SAS area, 120000 m^2 of building space was to be accommodated, and development was to proceed by stages. Older buildings adjoining the new development were to be refurbished, altered or demolished. SAS's new head office was subsequently to be dealt with in a separate detailed plan, whereas a detailed plan for the adjoining and evacuated barrack area would be deferred until the use of the barracks could be agreed.

Detailed plan and plan revision (August 1985 and October 1986)

The municipality presented a detailed plan as a purely formal precondition for building permission and to assure the general public of access to the water's edge of Brunnsviken and to other areas of attractive scenery. During the formal process of detailed planning, several consultations took place with the appropriate authorities. Exhibitions and evening briefings were put on for the general public.

The detailed plan regulated, in detail, the use to be made of different areas of land, allocating land for offices, parking, streets, footpaths and grounds. One area was delineated as a reserve for existing buildings of historic interest. In addition, land was earmarked for underground conduits. The maximum total area of buildings was fixed and it was decided within the detailed plan that the property could not be subdivided into different plots. In that way the ownership would continue to be in one hand. In addition, maximum heights were fixed for the various buildings. Next to the main building an area was designated for construction below ground. The reason for these detailed provisions was that the detailed plan was to form the basis of subsequent building permission. Construction of the head office had to be made possible, and steps had to be taken to forestall an application for a building permit for a building that did not tie in with the winning entry.

The SAS expansion plans also included a future second stage of 50000 m^2 building space that, in the winning entry, had been sited immediately southeast of the main building. Both in the competition assessment and in the compilation of a detailed plan, it was observed that this stage had been insufficiently studied, and so this part was excluded from the detailed planning. Instead any subsequent stage of expansion was referred back for renewed detailed planning work.

Before the new Planning and Building Act came into force in 1987, all detailed plans had to be confirmed by the county administration. In its confirmation resolution, the county administration excluded a traffic area provided for access to the E4 motorway, the reason being that the National Road Administration objected to the design and the location. The question was studied more closely and one year later, in October 1986, a new detailed plan was put forward with a satisfactory traffic arrangement.

Agreements concerning purchase and development (August 1985)

In August 1985 the parties concluded a combined purchase and development agreement that confirmed the previous agreement of principle.

This agreement also included an awkward and subsequently persistent point of uncertainty, in that the municipality undertook to draw up a detailed plan for the second phase of extension and to work for the confirmation of this detailed plan. Judging by this agreement, the parties took the view that the comprehensive plan that had been adopted should give SAS the requisite guarantees of continuing development in the area. This was conditional on SAS being able to show how the building development could be integrated with the landscape.

Several points in the previous agreement of principle were now enlarged on. SAS was to grant the municipality an easement so that it would be able, at no extra cost, to lay water and sewerage mains, power cables and district heating pipes within the SAS property. In addition, the general public were to be freely entitled to use the grounds of the property and its roads were to be open to the general public as pedestrians and cyclists. SAS was to construct, finance and in future maintain these roads. When the company had discharged all its obligations under the agreement, it would be deemed to have paid for municipal street expansion outside the property.

The agreement laid down that the location of the head office in this sensitive area was a one-off measure, and that accordingly the parties were agreed that the purchased property in its entirety should be used for the intended purpose and that it was not to change hands without municipal permission. SAS would be allowed, however, without consent, to transfer part or all of the property to companies within the SAS Group and also to a company in which SAS had a substantial proprietary interest. This clause was probably written in because the political leadership at the time wanted to see the property not as a tradable commodity but as exclusively dedicated to SAS.

Parcelling of the property took place in accordance with the purchase and development agreement. It comprised the detailed planning area and the unplanned area for expansion in phase two. The parcelled area comprised just under 27ha, of which 6ha was water.

Co-ordinated adoption

The area development plan, the detailed plan and the purchase and development agreement were adopted by the municipality on 26 August 1985. The revised detailed plan, regulating the access question, was adopted a year later. The final deed of sale, confirming payment of the purchase price, was also made out a year later.

Building work (November 1985 to January 1988)

The architects' competition was settled in January 1985, and in November that year the first bulldozers moved in. Excavating permission had been granted in October and building permission came in March 1986. Thus, building operations did not have to await the solution of the traffic question. Commissioning started in December 1987 and the official opening took place a month later. The office, intended for 1400 employees, was one of the largest construction projects in

Scandinavia at the time, with a floor space of 63000 m^2. One interesting detail is that the water in the gravel ridge underneath is used for both cooling and heating. Cold water is pumped up in summertime and heated by the building before being pumped down into another well where the temperature is about 10 degrees higher. During winter the warm water is pumped up to heat the building.

The construction cost came to SEK425 million, making about SEK6500 per m^2. The project as a whole cost about SEK650 million, which SAS provided out of equity. SAS never built the computer centre that was to have been located a little to the north of the head office, nor did it construct a separate building for training facilities. Because of a change in circumstances, these were no longer needed.

SAS had appointed the contractors Åke Larson Byggare AB (ÅLB) as its partner in directing building operations. ÅLB carried out planning work, procured contracts and materials and took charge of day-to-day building operations. ÅLB worked with a form of contracting of its own invention, based on close co-operation between client and ÅLB. Together with the client, the company advanced step by step until design, quality, function, production technology, timetable and permissible cost could all be determined with a calculating margin of 1–2%. This done, the expenditure programme and other requirements were finalized.

Adherence to the programmes became ÅLB's responsibility. Tenders were invited for subcontracts and for deliveries of goods and services as building work progressed. ÅLB then assessed the tenders for each subcontract together with the client, SAS. If any subcontract overshot the budget, then, conjointly with SAS, standard requirements had to be reduced when negotiating the next subcontract. With this procedure the client could be sure of the ultimate cost being on budget and that it would itself obtain maximum value for its money. Presumably this special form of contract facilitated the choice of new materials and structures of mainly good quality, without expenditure per square metre being greater than was normal for a more ordinary building. The large volume of building helped to achieve the same effect. Later on, however, there were leaking glass roofs and cracked façade tiling to be put right.

An old officers' mess was turned into a project office where ÅLB and SAS worked together. This gave SAS direct insight into everything that happened. As a consequence of this form of purchasing, a host of different contractors needed to be co-ordinated and issued with site information. All in all there were more than 150 subcontracts to be placed. There were about 500 construction workers on the move all the time, and with a variety of construction enterprises involved, personnel turnover was high.

End of the detailed plan implementation period

Such were the regulatory provisions of the 1987 Planning and Building Act, that the implementation period for the detailed plan expired in July 1992. SAS therefore activated the building permit on the last day of June, so as not to risk losing the existing, unutilized building rights of 15000 m^2. The building permit documents, however, were rather cursory because they did not contain what was required and, moreover, were partly at variance with the current detailed plan.

SAS withdrew its building permit application four months later, while stressing its great need for the continuing option of future expansion within the property.

SAS, however, transferred some of its activities elsewhere, including Arlanda airport, and now has more premises than it needs. This change of circumstances may lead the company to let a large part of its premises in the next few years. A possible sell-off of some parts of the operation could have the same effect. Consequently there is unlikely to be any need for an expansion of the Frösundavik facilities within the foreseeable future.

In 1993 it was rumoured that SAS intended selling the Frösundavik office complex and that the unconfirmed building rights of 15000 m^2 for the second phase were being discussed as part of the sale. SAS, however, informed the municipality that it had no such plans. On the other hand the company told the municipality in a letter that its discussions concerning co-operation with other European airlines could have implications for its property holding at Frösundavik.

Comprehensive planning programme 1992–3

In addition to the building rights that SAS had obtained in the 1985 detailed plan, there was the statement in the development agreement from the same point in time that the municipality pledged itself to work to obtain another 50000 m^2 of building rights within the SAS property to the east of the head office, within the part of the SAS property for which no detailed plan had been adopted (see Fig. 6.8).

However, the importance of the stretch of water and greenery extending through the municipality of Solna and a long way into Stockholm had increased with the passing years, and so the government, under powers conferred by the National Resources Act, requested from Solna and Stockholm an account of their plans for nature conservation, heritage conservation and outdoor life in this extensive area. The municipalities in turn drew up a joint programme for the future planning of the area.

In January 1993 the Solna municipal council resolved to join the City of Stockholm in submitting the programme to the county administration in response to the government's injunction. In that programme it was proposed that certain previously planned construction and development projects should be deleted. Among other things, the extension of the Frösundavik office complex should not be carried out because the SAS office complex and the expanded road system detracted from the natural unity of the area. Instead the green areas should be reinforced.

The municipal council also appended to the programme a supplementary statement that, briefly, went as follows. The present SAS head office comprises about 50000 m^2. In addition SAS holds nearly 20000 m^2 of unutilized building rights within the current detailed plan. Furthermore, the municipality of Solna realizes that it has pledged itself to work for the creation of building rights of about 50000 m^2 in the area immediately east of the existing buildings. The agreement, however, makes this conditional on SAS being able to show that it requires further space for head office or other purposes, and on it being established that the building development will not encroach on the surrounding environment.

According to municipal studies, however, building development on this scale cannot be integrated with the environment without excessive encroachment. Nor has SAS been able to establish any additional requirements for head office or other purposes. The municipality of Solna is also of the opinion that it was the State and SAS that asserted the necessity for expansion opportunities during the negotiations in the mid-1980s, whereas the municipality asserted environmental interests. The municipality of Solna therefore feels that the State should actively participate in discussions and negotiations between Solna and SAS concerning amendments to the agreement and detailed plan, with a view to making more adequate provision for nature conservation interests in the area.

For the long-term protection of the Ulriksdal–Haga–Brunnsviken–Djurgården area, the government decided, in the spring of 1994, to introduce an amendment to the Natural Resources (Conservation) Act, laying down that parks and historic landscapes within the area may not be altered by development enterprises or subjected to other kinds of interference impairing the cultural, natural or recreational value of the area. The amendment did not appear to be contentious, and was therefore likely to be passed by the Riksdag in the autumn.

Concluding reflections

The above description shows that the main outlines of what was agreed at the beginning of the development process concerning the actual head office were put into effect, but a few reflections of a more general nature are possible:

- It was the SAS project that initiated the planning procedures, not the other way around. This is the opposite procedure to that often presumed to apply to Swedish planning.
- A person buying a property from a municipality cannot be sure of the municipality's ability, or perhaps even its desire, to honour possible assurances concerning future building rights.
- In fact the property owner cannot even be sure of being allowed to use the unutilized building rights existing within an adopted detailed plan after the expiry of the implementation period. This means that the property owner cannot really obtain guarantees of a firm long-term right of future expansion. On the other hand, the owner is, in principle, quite certain of being allowed to exercise expansion opportunities on his property in accordance with the provisions of a detailed plan during the plan's implementation period.
- The State too is unable to abide by any assurances given on the subject of land use.
- Points of uncertainty in a development agreement can later result in awkward negotiations and unpredictable situations for the parties involved.
- So rapid are the changes occurring in business enterprise that a municipality cannot be certain of an apparently stable enterprise retaining all or part of its activities within a property, even for a short time.

6.5 The sale of freehold and tenant-owner properties

Swedish law is designed to enable the parties to do their own conveyancing without assistance from any intermediaries. All they need do is jointly sign a deed of purchase stating the designation of the property and the purchase price and declaring that the property is changing hands. A possession date is usually stated as well, failing which the possession date will be the date of the contract. Registration of title has to be applied for at the land registration authority, in order to publicize the new owner, who thereby becomes fully seized of the property. Often the deed of purchase is preceded by the signing of a contract of sale, stating the terms of the transaction in more detail. Standard forms of conveyancing are obtainable from handbooks.

Because in many cases property transactions are complicated, both sellers and buyers often choose to engage an estate agent. This is perhaps commonest in the conveyancing of single-family dwellings because normally these transactions involve non-professional, uninitiated parties. Rental and industrial properties are probably less frequently conveyed through estate agents, and the same goes for second homes. Another reason for engaging estate agents is that they can be expected to command a better view of the property market than individual persons. This is perhaps above all true in big-city markets, whereas in medium and small communities private individuals may have a close knowledge of the local property market.

Three examples of actual purchases of large properties will be given. These instances are based on interviews with estate agents but have for the most part been de-identified. In addition, a description will be given of the routines of a large estate agent for the conveyancing of single-family dwellings and tenant-owner properties.

Case 1: rental property in central Stockholm

The seller got in touch with the estate agent to commission him to sell a property consisting of rented apartments. A fixed-term contract was signed for the sale assignment. The estate agent carried out a valuation, compiled a four-page prospectus with facts about the property and advertised it in the daily papers. The prospectus gave an asking price of SEK6 million, which was SEK1 million less than the valuation. About 40 interested parties asked for the prospectus, and subsequently about ten of them inspected the property together with the estate agent. One interested party went further and obtained particulars about tenants and a list of rents. Housing tenancy agreements are fairly standardized and therefore did not need to be studied in detail.

When a property changes hands, responsibility for mortgage loans does not automatically pass to the new owner, even if the property still constitutes security for the loan. The lender indicated, however, that the intended purchaser would be allowed to take over the existing loans. Otherwise, purchaser would have had to look for a new mortgagee, possibly with the assistance of the estate agent.

The estate agent drew up a five-page contract of sale and, later, a one-page deed of purchase. The contract stipulated that the purchaser was to request an acquisition permit. Permits of this kind are necessary for the acquisition of rental properties in certain localities, the purpose being to keep unscrupulous landlords out of the market. The purchase price was rather less than SEK5 million. The estate agent's commission was 3% plus value added tax, which is a fairly normal amount for transactions involving less than SEK20–25 million. Larger transactions carry a lower percentage rate.

Case 2: rental property in central Stockholm

The seller commissioned the estate agent to sell a rental property in an exclusive location for at least SEK18 million. The estate agent took the same steps as in case 1 but also obtained technical particulars about the property from a technical expert. Close technical inspections are becoming increasingly common, and are normally paid for by the potential buyer. In this particular case, the property was found to be in need of refurbishment. The prospectus for the property was requested by 125 interested parties, and 15 were shown around before one of them signed a contract of sale. The tenants of the property, however, had previously formed a tenant-owner association (without any property holding) and notified the title registration authority that they wished to have first refusal of the property if it was put up for sale by the previous owner. Although the tenants were interested in purchasing the property, they did not wish to figure as speculators and in this way put up the price. Instead they awaited the sale to see whether the purchase price would be within their means. Otherwise, it is a rule of thumb that the price will be about 20% higher if the seller manages to sell the property to a tenant-owner association rather than to some other purchaser.

The purchaser must now wait and see whether the association wishes to take over the property on the terms of his own acquisition of it. Meanwhile, he has paid a deposit equalling 5% of the purchase price of not quite SEK20 million and is not really in a position to buy another property, because he does not know whether his present purchase will be completed. The purchaser cannot take possession of the property until it is made clear that the tenant-owner association does not wish to take his place. And he has to get a permit for the acquisition of rental property. At the time of writing, *the situation had not been resolved and the purchase had not been completed.*

Case 3: commercial property in a county town

The owner of a commercial property in a large county town commissioned the estate agent to dispose of it. The property, in an excellent central location, was in very good condition. The rentable floor space was just over 100000 m^2, one quarter of which was untenanted. The rent roll was estimated at SEK8 million annually. The vendor also wished to be relieved of management of the property until it was sold, and this task was entrusted to a company with which the estate agent was closely connected. A project group formed to conduct the sale included one representative each of the owner, the manager and the estate agent. A fairly comprehensive prospectus was compiled and the property advertised.

There were 15 requests for the prospectus, five or more of them coming from active prospects. The vendor, a credit company, held all the mortgage deeds and had previously acquired the property at a compulsory auction in order to cover its claims. The property was later sold for SEK60 million, of which SEK55 million was paid in the form of an instrument of debt to the seller with mortgage security. The estate agent's commission was 3%.

Cases 4 and 5 deal with an estate agent's routines for transactions involving single-family dwellings and tenant-owner properties. Although these descriptions apply to one organization, they probably convey an accurate picture of the working methods and business conditions of most estate agents in Sweden.

Case 4: detached and terraced houses

Nearly all sales of single-family dwellings begin with the vendor getting in touch with the estate agent. A contract is signed for the assignment, and this usually gives the estate agent the sole right of handling the transaction for three months. The minimum price for the property is fixed at the same time. The estate agent inspects the house and carries out a rough technical inspection. Together with the vendor the estate agent decides a reasonable asking price, in the light of his or her knowledge of the market. The house is advertised in the daily papers and any prospective buyers will be shown around by the estate agent. Prospective buyers requiring a closer technical inspection will normally have to arrange for it themselves. The buyer, if there is one, signs a contract of sale with the vendor and usually pays 10% of the purchase price cash down. The remainder is paid on the day of signing a definitive deal of purchase. The contract of sale is normally a standardized one of four pages and the deed of purchase, also standardized, a one-page document.

Before the contract of sale is signed, mortgage loans on the property are carefully examined with regard to size, interest rate, instalments and maturity date. The reason for inspecting these loans so carefully is that they will usually be taken over by the new owner; failing this, the loans have to be paid off, in which case the creditor returns the mortgage deeds to the property owner. The owner then has to surrender the mortgage deeds to the purchaser when the deed of purchase is signed, after which the purchaser can mortgage them elsewhere. If the loans are to be taken over, the usual thing is for the final deed of purchase to be signed at the bank concerned, in the presence of vendor, purchaser, estate agent and bank official. The reason for the deed of purchase being signed and the purchase price then transferred to the bank is that the bank takes over the documents of sale, armed with which it immediately applies for registration of the buyer's title. In this way the bank makes sure of the new owner quickly becoming responsible for both property and loans. The estate agent interviewed usually charges a commission equalling 4% of the purchase price, plus VAT. He can also arrange contacts with creditors if the buyer needs additional credits. Secondary homes are bought and sold on the same lines as permanent homes.

Case 5: tenant-owner properties

Normally it is the vendor who gets in touch with the firm of estate agents, and a contract is signed resembling that described in case 4. The estate agent investigates the flat and phones or writes to a representative of the tenant-owner association committee for a review of the association's finances. If possible the estate agent also finds out about the association's investment plans. An assessment can then be made of future monthly charges for the flat. After all these points have been covered, vendor and estate agent discuss a reasonable asking price, the estate agent drawing on experience of similar transactions. A small advertisement for the flat is published in the daily papers and put up in several bank windows. Prospective buyers are shown around by the estate agent. When the contract of sale is signed, 10% is normally paid cash down and the remainder on taking possession. The contract of sale is normally a three-page standard contract.

The tenant-owner association is liable for mortgage loans on the property, and so these are unaffected by a sale. In addition, the tenant-owner may have private loans on the security of the title. It is the duty of the tenant-owner association to keep a register of these private loans, and the normal thing is for the purchaser to take over any private loans on the same terms as the vendor. In cases where the purchaser needs new loans, the estate agent can also arrange contacts with credit agencies for this purpose.

The tenant-owner association has to approve the purchaser's admission to membership, and so the purchaser has to request admission to the association and the vendor has to be released from it. At the same time, the association is informed of the purchase price and other details. The membership application does not normally give rise to any problems, but as a precaution the contract of sale always includes a clause stipulating that, if membership is not obtained, the purchase will be cancelled.

The estate agent interviewed receives a commission of 5% plus VAT, subject to a minimum of SEK20000 with VAT included. It is worth adding that a share in a tenant-owner association is regarded as personal property. Documents of sale, consequently, are not public domain, unlike documents relating to the purchase and sale of real property.

Finally we may add that a person disposing of a tenant-owner title need not engage an estate agent at all. In that case the tasks otherwise performed by the estate agent devolve on the vendor.

CHAPTER 7
Concluding remarks

First of all one must say that it is difficult, perhaps even premature, to evaluate both the latest changes in urban development and the uneasiness following in the wake of the property crisis. This chapter, accordingly, is best regarded as a rather personal view of developments.

The central importance of public control for orderly urban development is not a new idea, but it was only after the Second World War, in connection with the rising aspirations of housing policy, that public planning measures achieved a wide-ranging impact. The aim was for each individual to have access to a good, spacious home at reasonable cost. Means employed to this end included detailed regulations concerning the qualities and design of homes. Another concern was with achieving functionally good building development on a large scale, for which collective, often municipal, solutions for the planning and management of housing development and local infrastructure were seen to be advantageous. Legislation, public administration and economic transfer systems were used as active instruments for achieving these policy aims. The motive force came from the serious housing shortage brought about by population movements and by demands for higher standards. Ambitions were also governed by political demands on the subject of social equality.

Although priority was given to questions of housing policy, there were other problems that demanded attention. Structural changes in enterprise created a need for new enterprise areas and town-centre renewals. As transport operations increased and the car became everyone's property, there were traffic problems to be solved. Urban development, then, was not just a matter of housing policy. Large parts of the physical structure of society had to be reconstructed.

Because the spate of building created opportunities for making large profits out of greenfield holdings, the government and municipalities joined together in pursuing an active land policy. Conditions for the award of State-subsidized housing loans were framed in such a way that land prices could be controlled. In addition, the municipalities often bought up land so as to have a reserve for future building development. Then again, rents were fixed with reference to the rental cost of apartments, mainly those of the non-profit municipal housing utilities. This made it difficult for other property owners to obtain a return on high land costs.

Social building development appears to have culminated in about 1970, when production of multi-family dwellings was at its peak. Up till then, there had on the whole been a consensus between the political parties on the framing of housing policy. Disagreements now began to grow more prominent. Government commissions affirmed, however, that the aims of housing policy should continue to focus on fairness and equality. But the widespread production, at the end of the 1970s, of single-family dwellings for freehold tenure showed that large groups of the population preferred more individual kinds of housing.

Following the downturn at the beginning of the 1980s and during the deregulation of the credit market, the conditions applying to urban development were transformed. Property prices rose, as did housing and non-residential production. There was political discussion as to whether public bodies were really coping with their tasks or were stick-in-the-muds without visions. With the credibility of public objectives tottering, however, political discussion seems to have become more frank compared with previous, more target-orientated periods. Previous axioms, such as the possibility of high-precision physical and social planning of society, have been called into question. It is too early, however, to say whether this is a passing doubt or a more permanent change of heart.

The past few years have been a turbulent time for Swedish ideas about planning. The relocation in 1989 of the national authority in charge of planning and housing questions to a part of southern Sweden with large regional problems can be said to betoken a toning-down of housing policy. Important changes also occurred after 1991, when a non-socialist government took office. That government closed down the Ministry of Housing and transferred its duties to six different ministries, thereby clearly indicating that planning and housing questions would not occupy as prominent a position in governmental and ministerial work as previously. The State has also curbed its position as financier, thereby depriving society of an important housing and land policy lever. In fact, public organization, financial supportive arrangements and legislation were remodelled so quickly that it is hard to sum up all the changes that have occurred in the past two to three years, let alone evaluate their consequences.

It is not necessarily the case, however, that urban development will continue to be deregulated and that the public sector should step back in favour of private solutions. Possibly the system has been cast in the role of a scapegoat, more sinned against than sinning. It is presumably too much to expect that planning and housing policy should cope with the rapidly changing conditions by which players in the property market were confronted during the deregulation of the credit market in the 1980s. Large additional credits entered the property market, resulting in rising property prices. Most of the credits went to the non-housing rental market, but they also helped to raise prices in other submarkets and to boost output of both housing and non-residential property, at heavy cost. Market forces, or perhaps rather the credit market, seemed to have grown too strong to be controlled by public policy.

During the ensuing recession, there was in a manner of speaking also a lack of planning tasks for national and local government, because low demand and the small volume of building output afforded no scope for planning and regulation. Thus, high demand and a large output of new homes and non-housing premises quickly gave way to low demand and a small volume of new output. Existing floor space was more than enough. Thus, the public organization of the time could not cope with the extreme situation prevailing during the boom period and had little to do during the recession. No wonder, then, that its activities were challenged. If this conjectural argument holds good, then planning will presumably enter a period of revival when the economy stabilizes and building gets going again. Environmental awareness in society, if nothing else, should contribute in this respect, because environmental issues demand a holistic view of changes in the physical structure of society. Presumably, though, the institutional frameworks of urban development will remain altered, like the balance of power between different interests.

As has been made clear, the municipalities play a very prominent part in urban planning and in the operation of local infrastructure. They have been reorganized, more or less radically, in recent years, a process partly made possible by amendments to local government legislation. As a result, certain municipal organizations may possibly have become indistinct in the minds of outsiders. In other words, it can be hard to tell who is responsible for what. Another result is that knowledge of the organizational structure in one municipality is not directly transferable to another, which can make it hard for the general public to get their bearings in the decision-making hierarchies. The freer principle of organization, on the other hand, should result in municipalities adopting a more flexible response to new situations.

Municipal activities in some fields are tending to change from outright political activity to more business-related operations. The incorporation of municipal activities, such as the formation of water and sewerage companies, is one indication of this. The aim, however, is to preserve or perhaps even augment control over urban development, even if day-to-day work is entrusted to others. There has also been a tendency for development activities to move in favour of responding to developers' initiatives, whereas previously these activities expressed a municipal intent supported by public policy instruments. The negotiations taking place in this connection between municipality and developers have often preceded the actual planning. It is too early to say whether much planning by negotiation was a boom phenomenon or represents a more permanent state of affairs.

The municipalities are also tending to a greater extent to source externally services that they used to provide themselves. The practical business of planning is a service of this kind. Although, of course, control still exists in the form of plan adoption and building permits, there is a danger that the municipalities that decide to put their planning work out to contract will miss out on the knowledge accruing from everyday planning work. This could have the effect of transforming the municipality from an active planning authority into a passive sanctioning authority. Another important change for housing policy concerns the non-profit housing utilities, which are closely bound up with municipal interests. Although these utilities have to operate on a non-profit basis, there are many indications that they are beginning to act in the housing market in the same way as private housing companies. At the same time, the non-profit utilities have lost their special position in the credit market: previously they received more favourable loans than others.

Developments in the property market during the 1980s and 1990s show that belief in the possibility of controlling developments through public planning was exaggerated by comparison with what was actually feasible. Post-war experience, however, suggests that social objectives have to be supplemented by planning and by control of finance, land holdings, planning instruments, and so on, so as to make sure that the objectives are achieved. If the end product is managed under one's own auspices, this naturally increases the possibility of control still further. On the other hand, the thoroughly regulated, rigidly controlled society is distrusted by large groups of the population, because it hardly helps to satisfy the varying preferences of a well educated, environmentally conscious and relatively affluent population.

It seems unclear how the interaction of private and public sectors is to be organized in the long term. Everyone involved, including the spokesmen of the main political parties, seems to agree that land and housing policy is in need of a further review. A commission has been appointed, for example, to review the system of planning and permits, the aim being to enhance the efficiency of the system, while ensuring that environmental interests can be provided for. Probably this will lead to a reinforcement of comprehensive planning at the expense of detailed planning. A continuing review of housing policy is also needed, because a housing shortage may develop within the near future, owing to the extremely low level of building output for some years past. The shortage may be aggravated by a low level of municipal planning and land preparedness preventing new projects materializing at the rate required by an upsurge of demand. Thus, there is a risk of the classical pattern being reiterated, i.e. of housing production lagging behind fluctuations in demand instead of keeping pace with them.

Those who work with physical planning or in the property business will perhaps find themselves acting in a world where political decisions are no longer as important as they used to be and are supplanted by the assessment of different projects by market forces and, above all perhaps, by financiers. Developments up until 1990, however, point to the perils of relying excessively on a deregulated market. Although what happened in Sweden was presumably an extreme case, there are certain lessons to be learned from it. Economic development impelled building interests with plenty of capital to produce a glut, mainly of non-housing premises but also of expensive apartments. Glut and economic recession led to widespread vacancies and falling property prices. Even the banking system was shaken. However, the State was able to resolve the situation by dint of very strong supportive measures. The need for an effective interaction between public and private players was clearly revealed.

BIBLIOGRAPHY

Affärsvärlden (1984 to 1993). Weekly magazine, Sweden.

Åström, K. 1993. *Stadsplanering i Sverige*. Stockholm: Byggförlaget.

Bankstödsnämnden 1993. *Directors report, May 1 - June 30, 1993*. Stockholm: Bank Support Authority/Bankstödsnämnden.

Bankstödsnämnden/Valuation Board 1994. *Värdering av fastigheter i bankstödsprocessen Slutrapport 1994.03.16*. Stockholm: Bank Support Authority/Bankstödsnämnden.

Bengtsson, B. & A. Victorin 1991. *Hyra och annan nyttjanderätt till fast egendom*. Stockholm: Norstedts Juridik.

Bernitz, U., L. Heuman, M. Löfmarck, H. Ragnemalm, P. Seipel, A. Victorin, H-H. Vogel 1991. *Finna rätt - Juristens källmaterial och arbetsmetoder*, 3rd edn. Stockholm: Juristförlaget.

Bjerkén, T. 1990. Expropriation in Sweden. In *Compensation for expropriation: a comparative study* (vol. I), G. M. Erasmus (ed.), 121–51. Oxford: The United Kingdom National Committee of Comparative Law.

Bladh, M. 1991. *Bostadsförsörjningen 1945–1985*. Report SB: 42, Statens Institut för byggnadsforskning, Gävle.

Bostadsdepartementet, Kommunförbundet, Lantmäteriverket, Planverket 1987. *Plan- och bygglagstiftningen*. Sweden: National Land Survey/Lantmäteriverket.

Bostadsstyrelsen & Lantmäteriverket 1988. *Markpolitiska medel för kommunerna*. Report 4, National Land Survey/Lantmäteriverket, Gävle.

Bouvin, Å. & B. Hedman 1972. *VA-lagstiftning*. Stockholm: Norstedts Gula Bibliotek.

Boverket 1991. Studie av markvillkoret och dess tillämpning. Paper Bo 5–8, SCB Reports Serie, National Board of Housing, Building and Planning/Boverket, Karlskrona.

— 1993a. *Informationsblad rörande nya regler för bostadsfinansiering*. National Board of Housing, Building and Planning/Boverket, Karlskrona.

— 1993b. *Svensk bostadsmarknad i internationell belysning*. Report 1993: 2, National Board of Housing, Building and Planning/Boverket, Karlskrona.

— 1994. *Bostadsmarknaden och 90-talets förändringar*. Report 1994: 1, National Board of Housing, Building and Planning/Boverket, Karlskrona.

Byggentreprenörerna 1994. *Fakta om byggandet*. Stockholm: Byggentreprenörerna.

Cars, G. 1992. *Förhandlingar mellan privata och offentliga aktörer i samhällsbyggandet*. Stockholm: Department of Regional Planning, Royal Institute of Technology/KTH.

Catella 1993. *Market report, 1993: 1*. Stockholm: Catella.

— 1994. *Market report, 1994: 1*. Stockholm: Catella.

CFD 1993. *Statistik över fastighetsbeståndet i län och kommuner*. Statistical data from the Central Board for Real Estate Data/Centralnämnden för fastighetsdata, Gävle.

Ds 1994. *Förslag till skydd för området Ulriksdal-Haga-Brunnsviken-Djurgården*. Ministerial memorandum/departementspromemoria 1994: 3, Ministry of the Environment and Natural Resources/Miljö- och naturresursdepartementet, Stockholm.

Ds Bo 1985. *Stadsförnyelse- en inventering av regler*. Ministerial memorandum/departementspromemoria 5, Bostadsdepartementet, Stockholm.
Didón, L. U., L. Magnusson, O. Millgård, S. Molander 1987. *Plan- och bygglagen: en kommentar*. Stockholm: Norstedts Gula Bibliotek.

Fastighetsvärlden 1993. *Svensk fastighetsindikator 1993. Fastighetsägarens årsbok*. Täby: Tidnings AB Fastighetsvärlden.
Forsberg, H. 1992. *En politisk nödvändighet. En studie av den fysiska riksplaneringens introduktion och tillämpning*. PhD thesis, Linköping Studies in Art and Science, University of Linköping.

Gavlefors, U. & H. Roos 1992. *Prisbildning på bostadsrätter*. Report 6, National Land Survey/Lantmäteriverket, Gävle.
Gustafsson, A. 1992. *Kommun och landsting idag*. Malmö: Gleerups.

Hägred, U. 1994. Vem flyttar? See Boverket (1994), 40–52.
Hall, B., B. Hansson, B. Ljungqvist, L. Magnusson, R. Strömgren 1988. *Fastighetsrättsliga ersättningsprinciper: handbok med beräkningsexempel*. Stockholm: Norstedts.
Hall, T. (ed.) 1991. *Planning and urban growth in the Nordic countries*. London: Spon.
Hållén, G., J. Lidvall, S. Lindgren 1982. *Småhusupphandling; förutsättningar för avtal*. Report R54, Council for Building Research/ Byggforskningsrådet, Stockholm.
Hemström, C. 1986. *Gemensamhetsanläggningar: inrättande och förvaltning*. Stockholm: Norstedts.
Holmberg, E. & N. Stjernqvist 1980. *Grundlagarna - med tillhörande författningar*. Stockholm: Norstedts.
Hornsved, G. 1991. *Exploateringsavtal för fritidsbebyggelse på privatägd mark (EFP 91)*. Stockholm: Swedish Association of Local Authorities/Kommunförbundet.
Hultin, O. 1985. Tävling om SAS huvudkontor. *Arkitektur* (3), 16–7.

Industriförbundet 1992. *Sveriges Industri*. Stockholm: Federation of Swedish Industries/ Industriförbundet.
Ingves, S. 1993. Bankkrisen en fastighetskris. *Lantmäteritidskriften* (6), 6–9.

Jacobsson, L. 1994. Handeln och fastighetsmarknaden. In *Fastighetsägande inför år 2000 - visioner och strategier*, SFF 41–55. Stockholm: The Swedish Federation for Rental Property Owners/Sveriges fastighetsägarförbund.
JM Byggnads och Fastighets AB 1993. *Bo som du vill på Sticklingehöjden*. Informationsbroschyr. Stockholm: JM Byggnads och Fastighets.

Kalbro, T. 1992. *Markexploatering: ekonomi, juridik, teknik och organisation*. Report 4, National Land Survey/Lantmäteriverket, Gävle.
— & H. Larsson 1983. *Tillämpning av anläggningslagen*. Report 4: 40, Department of Real Estate Planning, Royal Institute of Technology/KTH, Stockholm.
Kommunförbundet 1982. *Gatukostnader: handledning vid uttag av anläggningskostnader för gator, parker och andra allmänna platser*. Stockholm: Swedish Association of Local Authorities/Kommunförbundet.
Konjunkturinstitutet 1993. *Konjunkturläget hösten 1993*. Stockholm: National Institute of Economic Research/Konjunkturinstitutet.

Larsson, H. 1988. *Om taxor och avgifter i samhällsbyggandet*. Report 4: 57, Department of Real Estate Planning, Royal Institute of Technology/KTH, Stockholm.
Lif, G. 1986. Tid är pengar-planering av SAS' huvudkontor i Solna. *Stadsbyggnad* (3), 16–8.
Liman, L-O. 1983. *Entreprenadrätt - en introduktion*. Stockholm: Svensk Byggtjänst.
Lindgren, E. 1992. Varför reglerar vi markanvändning och byggande? *Lantmäteritidskriften* (1), 5–12.
LMV 1988a. *Boken om genomförande enligt plan- och bygglagen*. Report 3, National Land Survey/Lantmäteriverket, Gävle.
— 1988b. *Genomförandebeskrivning till detaljplan*. Report 25, National Land Survey/Lantmäteriverket, Gävle.
— 1989. *Utvärdering av ekonomiska konsekvenser vid planering och genomförande av detaljplaner - metoder*. Report 6, National Land Survey/Lantmäteriverket, Gävle.
— 1991. *Handbok i skattefrågor vid fastighetsbildning*. National Land Survey/Lantmäteriverket, Gävle.
— 1993a. *Fastighetsbarometern*. National Land Survey/Lantmäteriverket, Gävle.
— 1993b. *Ersättningshandboken*. National Land Survey/Lantmäteriverket, Gävle.
— 1993c. *Planvinstbegreppet*. National Land Survey/Lantmäteriverket, Gävle.
— 1993d. *Fastighetsmarknaden idag*. Report 12, National Land Survey/Lantmäteriverket, Gävle.
— 1994a. *Fastighetsbarometern*. National Land Survey/Lantmäteriverket, Gävle.
— 1994b. *Fastighetmarknaden idag*. Report 15, National Land Survey/Lantmäteriverket, Gävle.
Larsson, G. 1979. *Fritidsboende och fritidsbebyggelse*. Report T23, Council for Building Research/Byggforskningsrådet, Stockholm.
— 1991. *Exploatering i samverkan: En jämförande internationell studie*. Report 4: 65, Department of Real Estate Planning, Royal Institute of Technology/KTH, Stockholm.
Lodin, S-O., G. Lindencrona, P. Melz, C. Silfverberg 1993. *Inkomstskatt*. Lund: Studentlitteratur.
Lundström, S. 1992. *Fastighetsföretagande*. Report 5: 30, Department of Real Estate Economy, Royal Institute of Technology/KTH, Stockholm.
Löfmarck, M. 1991. Förarbeten. See Bernitz et al. (1991), 87–108.

Mattsson, H. 1992. Sweden. See Wood & Williams (1992), 193–226.
Michanek, G. 1989. *Lagbok i miljörätt 1989*. Stockholm: Allmänna.
Miller, T. 1993. *Genomförandeavtal i exploateringsprocessen med särskild inriktning mot föravtal*. Karlskrona: National Board of Housing, Building and Planning/Boverket and Swedish Association of Local Authorities/Kommunförbundet.
Ministry of Finance 1992. *The Swedish budget, 1992/93*. Stockholm: Ministry of Finance/Finansdepartementet.

Nationalencyklopedin 1989– . Sweden: Bra Böckers.
Nyström, J. 1994. *Tätortslandskapet i Sverige*. Report 1: 94, Department of Human Geography, Stockholm University, Stockholm.

Österberg, T. 1990. *Samfälligheter: handbok för samfällighetsföreningar*. Stockholm: Allmänna.

Planverket 1987a. *Boken om detaljplan och områdesbestämmelser*. Stockholm: Planverket.

— 1987b. *Boken om översiktsplan*. Stockholm: Planverket.
— 1988. *Boken om lov - planmässig prövning*. Stockholm: Planverket.
von Platen, F. 1991. Allt är inte förhandlingsplanering. *Planera, bygga, bo*. (2), 3.
Prop. 1985/86: 1. *Ny plan- och bygglag*. Stockholm: Regeringens proposition.
— 1992/93: 150. *Regeringens proposition med förslag till slutlig reglering av statsbudgeten för budgetåret 1993/94*. Stockholm: Regeringens proposition.

Redlund, M. 1987. Vi-känsla ger SAS-huset extra kvalitet. *Byggindustrin* (2), 22–3.
— 1988. SAS-kontor i buisness class till lågpris. *Byggindustrin* (8), 11–5.
Råckle, G. 1993. *Sveriges småhusmarknad - en blivande krisbransch?* Professional Paper, Department of Real Estate and Construction Management, Royal Institute of Technology/KTH, Stockholm.
— 1994. *Alternativa finansieringslösningar för fastigheter*. Report 5: 37, Department of Real Estate and Construction Management, Royal Institute of Technology/KTH, Stockholm.

Sandblad, J. 1984. *Plangenomförande: Kommunen som avtalspart, markföretagare och förnyelsemotor*. Report R126, Council for Building Research/Byggforskningsrådet, Stockholm.
SCB various years. *Reports Serie AM, Bo, K, N, Na, P and R*. Statistics Sweden/Statistiska Centralbyrån, Stockholm.
— various years. *Folk- och bostadsräkningarna 1960, 1980 och 1990 (Population and housing censuses 1960, 1980 and 1990)*. Stockholm: Statistics Sweden/Statistiska Centralbyrån.
— 1980. *Bostads- och byggnadsstatistisk årsbok 1980*. Stockholm: Statistics Sweden/Statistiska Centralbyrån.
— 1985–93. *Statistical yearbook*. Stockholm: Statistics Sweden/Statistiska Centralbyrån.
— 1986–93. *Fastighetsprisstatistik*. Reports P18SM8601–P18SM9301, Statistics Sweden/Statistiska Centralbyrån, Stockholm.
— 1990a. *Rikets fastigheter 1989, del 1*. Report BO37SM9001, Statistics Sweden/Statistiska Centralbyrån, Stockholm.
— 1990b. *Rikets fastigheter 1988, del 2*. Report BO38SM8901, Statistics Sweden/Statistiska Centralbyrån, Stockholm.
— 1990c. *Naturmiljön i siffror*, 3rd edn. Stockholm: Statistics Sweden/Statistiska Centralbyrån.
— 1991a. *Enskilda försäkringsbolag 1989*. Stockholm: Statistics Sweden/Statistiska Centralbyrån.
— 1991b. *Rikets fastigheter 1990, del 2. Småhus*. Report BO38SM9101, Statistics Sweden/Statistiska Centralbyrån, Stockholm.
— 1991c. *Tema invandring*. Stockholm: Statistics Sweden/Statistiska Centralbyrån.
— 1992a. *Bostads- och byggnadsstatistisk årsbok 1992*. Stockholm: Statistics Sweden/Statistiska Centralbyrån.
— 1992b. *Fastighetsprisstatistik 1991; jord- och skogsbruk*. Report P20SM9201, Statistics Sweden/Statistiska Centralbyrån, Stockholm.
— 1992c. *Folk- och bostadsräkningen 1990, del 2: Folkmängd och sammanboende*. Stockholm: Statistics Sweden/Statistiska Centralbyrån.
— 1992d. *Folk- och bostadsräkningen 1990, del 3: Lägenheter*. Stockholm: Statistics Sweden/Statistiska Centralbyrån.
— 1992e. *Folk- och bostadsräkningen 1990, del 4: Hushåll*. Stockholm: Statistics Swe-

den/Statistiska Centralbyrån.

— 1992f. *Folk- och bostadsräkningen 1990, del 5: Förvärvsarbete och yrke*. Stockholm: Statistics Sweden/Statistiska Centralbyrån.

— 1992g. *Markanvändningen i tätorter 1990 och förändringen 1980–1990*. Report Na14SM9201, Statistics Sweden/Statistiska Centralbyrån, Stockholm.

— 1992h. *National wealth and stocks of fixed assets 1980–1990*. Stockholm: Statistics Sweden/Statistiska Centralbyrån.

— 1992i. *Rikets fastigheter, del 1*. Report Bo37SM9101, Statistics Sweden/Statistiska Centralbyrån, Stockholm.

— 1992j. *Tätorter 1990*. Report Na38SM9201, Statistics Sweden/Statistiska Centralbyrån, Stockholm.

— 1993a. *Bostads- och byggnadsstatistisk årsbok 1993*. Stockholm: StatisticsSweden/Statistiska Centralbyrån.

— 1993b. *Bostads- och hyresundersökningen 1992*. Report Bo31SM9301, Statistics Sweden/Statistiska Centralbyrån, Stockholm.

— 1993c. *Byggnadskostnadsstatistik 1992*. Report Bo22SM9303, Statistics Sweden/Statistiska Centralbyrån, Stockholm.

— 1993d. *Fastighetsprisstatistik 1992; jord- och skogsbruk*. Report P20SM9301, Statistics Sweden/Statistiska Centralbyrån, Stockholm.

— 1993e. *Fastighetsprisstatistik 1992; småhus, hyreshus, industrifastigheter och obebyggd tomtmark*. Report P18SM9301, Statistics Sweden/Statistiska Centralbyrån, Stockholm.

— 1993f. *Finansiella företag 1991*. Stockholm: Statistics Sweden/Statistiska Centralbyrån.

— 1993g. *Jordbruksstatistisk årsbok*. Stockholm: Statistics Sweden/Statistiska Centralbyrån.

— 1993h. *Markanvändningen i Sverige*, 2nd edn. Stockholm: Statistics Sweden/Statistiska Centralbyrån.

— 1993i. *Rikets fastigheter 1992, del 1*. Report Bo37SM9201, Statistics Sweden/Statistiska Centralbyrån, Stockholm.

— 1993j. *Rikets fastigheter 1993*. Report Bo37SM9301, Statistics Sweden/Statistiska Centralbyrån, Stockholm.

— 1993k. *Skogsstatistisk årsbok*. Stockholm: Statistics Sweden/Statistiska Centralbyrån.

— 1993m. *Tätortsexpansion på jordbruksmark 1960–1990*. ReportNa10SM9301, Statistics Sweden/Statistiska Centralbyrån, Stockholm.

— 1993n. *Småhusbarometern*. Report 263, Statistics Sweden/Statistiska Centralbyrån, Stockholm.

— 1994a. *Bostadsbyggandet: nybyggnad och modernisering 1a–4e kvartalet 1992*. Report Bo14SM9401, Statistics Sweden/Statistiska Centralbyrån, Stockholm.

— 1994b. *Nybyggnadskostnader för gruppbyggda småhus och flerbostadshus 1993*. Report Bo22SM9401, Statistics Sweden/Statistiska Centralbyrån, Stockholm.

— 1994c. *Outhyrda lägenheter i flerbostadshus den 1 mars 1994*. Report Bo34SM9401, Statistics Sweden/Statistiska Centralbyrån, Stockholm.

SFF 1994. *Real estate investments in Sweden*. Stockholm: Swedish Federation for Rental Property Owners/Sveriges fastighetsägarförbund.

SML 1990. *Lokalhyresmarknaden i Stockholm*. Report 1990: 1, Stockholms Mark- och lokaliseringsbolag, Stockholm.

— 1993. *Lokalhyresmarknaden i Stockholm*. Report 1993: 1, Stockholms Mark- och lokaliseringsbolag, Stockholm.

SNA 1991. *The national atlas of Sweden: the population*. Sweden: Bra Böcker.
Söderberg, J. 1985. *Att upphandla byggprojekt*. Lund: Studentlitteratur.
Solna and Stockholm 1992. *Program för planering av Ulriksdal-Haga-Brunnsviken-Djurgården 1992*. Solna: Solna Kommun.
SOU 1945. *Slutbetänkande avgivet av Bostadssociala utredningen. Del 1. Allmänna riktlinjer för den framtida bostadspolitiken* (1945: 63). Stockholm: Statens Offentliga Utredningar.
— 1974a. *Solidarisk bostadspolitik* (1974: 17). Stockholm: Statens Offentliga Utredningar.
— 1974b. *Markanvändning och byggande. Principer för lagstiftning* (1974: 21). Stockholm: Statens Offentliga Utredningar.
— 1979. *Hushållning med mark och vatten 2. Del II. Bakgrundsbeskrivning* (1979: 55). Stockholm: Statens Offentliga Utredningar.
— 1984a. *Bankrörelselag* (1984: 26). Stockholm: Statens Offentliga Utredningar.
— 1984b. *Bostadskommittens delbetänkande. Sammanfattning* (1984: 34). Stockholm: Statens Offentliga Utredningar.
— 1988. *Förnyelse av kreditmarknaden* (1988: 29). Stockholm: Statens Offentliga Utredningar.
— 1989. *Hypoteksinstituten i framtiden* (1989: 103). Stockholm: Statens Offentliga Utredningar.
— 1990a. *Privat tjänstesektor* (1990: 14). In Appendix 17 to Långtidsutredningen 1990. Stockholm: Statens Offentliga Utredningar.
— 1990b. *Tomträttsavgäld* (1990: 23). Stockholm: Statens Offentliga Utredningar.
— 1990c. *Konkurrensen inom bygg/bosektorn* (1990: 62). Stockholm: Statens Offentliga Utredningar.
— 1991. *Fastighetsleasing. Sale- and lease back* (1991: 81). Stockholm: Statens Offentliga Utredningar.
— 1992. *Långtidsutredningen 1992* (1992: 19). Stockholm: Statens Offentliga Utredningar.
— 1993a. *Stabilisering av bostadskreditmarknaden* (1993: 14). Stockholm: Statens Offentliga Utredningar.
— 1993b. *Nya villkor för ekonomi och politik* (1993: 16). Stockholm: Statens Offentliga Utredningar.
— 1993c. *Miljöbalk* (1993: 27). Stockholm: Statens Offentliga Utredningar.
— 1994. *Följdlagstiftning till miljöbalken* (1994: 96). Stockholm: Statens Offentliga Utredningar.
Stockholms kommun 1990. *Nya planprocessen: Riktlinjer för detaljplanearbetet*. Kommunstyrelsen utlåtande 1990: 334, antaget av kommunfullmäktige 1990-12-17, Stockholm City/Stockholm stad, Stockholm.
STR 1993. *Trähusbranschen*. Sweden: Sveriges Trähusfabrikanters Riksförbund.
Sundqvist, A. 1987. Ny fastighetstaxering- de regionala skillnaderna ökar. *Svensk lantmäteritidskrift* (3), 27–8.
Svedinger, B. 1989. *Stadens tekniska infrastruktur: En kunskapsöversikt*. Stockholm: Report T4: 1989, Council for Building Research/Byggforskningsrådet.
Svensk Byggtjänst 1994. *Fastighetsnomenklatur*. Solna: Svensk Byggtjänst.
Svenska Vatten- och Avloppsverksföreningen 1982. *Kommunal VA-taxa*. Publikation P44 from Svenska Vatten- och Avloppsverksföreningen, Stockholm.
Sveriges Rikes Lag 1994. *Sveriges Rikes lag*. Stockholm: Norstedts.
Swedish Institute 1991. *The Swedish economy: facts about Sweden*. Stockholm: Swedish

Institute/Svenska Institutet.

USK 1992. *Statistisk årsbok för Stockholm 1993*. Stockholm: Stockholm City/Stockholm stad.

Valuation Board 1994. *Valuation guidelines: makroekonomiska faktorer och fastighetsekonomiska nyckelparametrar*. Stockholm: Valuation Board.

Vedung, E. 1993. *Statens markpolitik, kommunerna och historiens ironi*. Stockholm: SNS.

Victorin, A. 1990. *Kommersiell hyresrätt*. Stockholm: Juristförlaget.

— 1993. *Bostadsrätt*. Stockholm: Juristförlaget.

Wallmark, J. 1991. *Detaljplaneringens inverkan på fastigheters marknadsvärden*. Report, Department of Real Estate Planning, Royal Institute of Technology/KTH, Stockholm.

Werin, L. 1978. Expropriation - en studie i lagstiftningsmotiv och ersättningsrättsliga grundprinciper. *Svensk Juristtidning* (6), 81–120.

Wilhelmsen, A. M. & H. Lindgren 1993. Landets industrihus i databas. *Byggforskning* (3/4), 16–8.

Wood B. & R. H. Williams (eds) 1992. *Industrial property markets in western Europe*. London: Spon.

Zingmark, A. 1994. Lokalmarknaden i Sverige - struktur och utvecklingstendenser. In *Fastighetsägande inför år 2000 - visioner och strategier*. Swedish Federation for Rental Property Owners/Sveriges fastighetsägarförbund, SFF, 27–40. Stockholm: SFF.

Index

For Product Safety Concerns and Information please contact our EU representative GPSR@taylorandfrancis.com
Taylor & Francis Verlag GmbH, Kaufingerstraße 24, 80331 München, Germany

www.ingramcontent.com/pod-product-compliance
Ingram Content Group UK Ltd.
Pitfield, Milton Keynes, MK11 3LW, UK
UKHW020203200926
475702UK00004B/591
* 9 7 8 1 1 3 8 4 9 0 1 3 0 *